The Open University

Mathematics and Computing
A first level multidisciplinary course

Open **Mathematics**

UNIT

9

BLOCK B
EVERY PICTURE TELLS A STORY

Music

Prepared by the course team

MU120 course units were produced by the following team:

Mike Crampin (Author)

Margaret Crowe (Course Manager)

Fergus Daly (Academic Editor)

Chris Dillon (Author)

Judy Ekins (Chair and Author)

Barrie Galpin (Author)

Alan Graham (Author)

Linda Hodgkinson (Author)

Gillian Iossif (Author)

Eric Love (Author and Academic Editor)

David Pimm (Author and main Academic Editor)

Other contributions to the text were made by a number of Open University staff and students and others acting as consultants, developmental testers, critical readers and writers of draft material. The course team are extremely grateful for their time and effort.

The course units were put into production by the following:

Course Materials Production Unit (Faculty of Mathematics and Computing)

Martin Brazier (Graphic Designer)

Hannah Brunt (Graphic Designer)

Alison Cadle (TEXOpS Manager)

Jenny Chalmers (Publishing Editor)

Sue Dobson (Graphic Artist)

Roger Lowry (Publishing Editor)

Diane Mole (Graphic Designer)

Nazlin Vohra (Graphic Designer)

The Open University, Walton Hall, Milton Keynes, MK7 6AA.

First published 1996. Reprinted 1997.

Edited, designed and typeset by the Open University using the Open University TEX System.

Printed in the United Kingdom by Henry Ling Ltd, The Dorset Press, Dorchester, Dorset DT1 1HD.

ISBN 0 7492 2235 2

This text forms part of an Open University First Level Course. If you would like a copy of *Studying with The Open University*, please write to the Central Enquiry Service, PO Box 200, The Open University, Walton Hall, Milton Keynes, MK7 6YZ. If you have not already enrolled on the Course and would like to buy this or other Open University material, please write to Open University Educational Enterprises Ltd, 12 Cofferidge Close, Stony Stratford, Milton Keynes, MK11 1BY, United Kingdom.

Contents

Study guide

This unit consists of five sections, each concerned with some aspect of mathematics and musical sound.

Music is created from sounds, so you may not be surprised to find that a substantial amount of the material for this unit is in the form of audiotape bands. Section 1 starts with an audiotape discussion between a musician and a mathematician. Both mathematics and music have their own technical vocabularies and some basic musical phenomena are described from these two perspectives. The remainder of this section examines the idea of musical scales. Section 2 looks at ways of describing, from a mathematical viewpoint, how much higher or lower one note is than another.

Section 3 makes use of a videotape band, but also draws on a short audiotape band in which the musician and mathematician discuss their impressions of this video in the light of their previous conversation.

There are three sections of the *Calculator Book* to study. The first two, in Section 4, explore the sine function (using the SIN button on your calculator) and the third, in Section 5, explores the mathematics of two different ways of specifying the notes in the basic Western musical scale.

At the end of Section 5, in the final audiotape band, a piano tuner describes and illustrates how he sets about tuning a piano.

There is an associated television programme, *A Language for Movement*, which looks at various forms of dance notation, and its links to musical notation. There are also two optional elements. The first of these is the appendix to this unit which gives a brief account of Western musical notation, viewing it as a particular sort of graph. This makes use of a piece of music written in the year 1720 which accompanies one of the dances shown in the television programme. There is also a short extract in the readings for Block B which you may wish to read after finishing this unit. It may give you insight into how far your understanding of the mathematics of music has progressed.

Because this unit draws on all of the learning media used in this course, you have an opportunity to revise and develop your ability to work with them. One of the main learning outcomes for this unit will be improving your ability to learn from audiotape bands.

In your work schedule, be sure to include time for completing an assessment question.

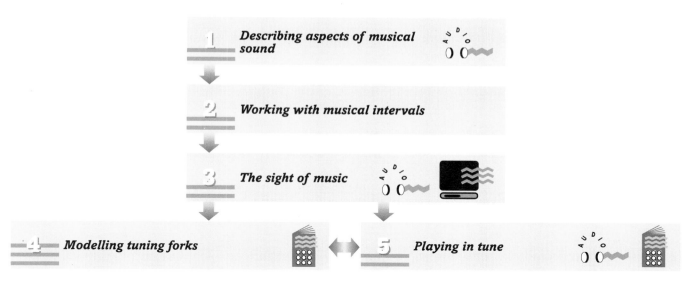

Summary of sections and other course components needed for *Unit 9*

Introduction

St. Cecilia is the patron saint of musicians: mathematicians, it seems, have to get by without one.

Blessed Cecilia, appear in visions
To all musicians, appear and inspire:
Translated Daughter, come down and startle
Composing mortals with immortal fire.

(*Hymn to St. Cecilia* by W. H. Auden, set to music by Benjamin Britten)

This unit concerns musical sound. The main underlying theme is using mathematics to explain and to help understand the world of musical sound and the construction of certain musical instruments. Mathematics is used to examine both some aspects of music and some ideas about acoustics, the physical study of sound. People have tried to explore music mathematically throughout history: from Pythagoras and Archytas dating from around 400 BC to twentieth-century composers such as Schönberg, Stockhausen or Xenakis.

First, a word of reassurance. You do not have to be 'musical' to work on the material of this unit, nor do you have to be able to read music or play an instrument. All of the terms used to talk about music are explained in the course of the unit. Even if you do have some of this background experience, you may well not have thought about music in this way before. It is the *mathematical* ideas and relationships that musical sound embodies which are most in focus.

These are some of the specific questions this unit examines.

◇ What is a musical pitch, what is a musical note, and how do different notes relate to one another?

◇ What makes up a musical scale, and why are there, at most, twelve 'different' notes in any Western musical scale?

◇ What is the function of frets found on some stringed instruments, and where are the frets on a guitar placed? Why are the distances between them not all the same?

Guitar

◇ What does it mean for an instrument to be 'in tune', and how and why are different instruments, such as a piano and a violin, tuned the way they are?

◇ How can a sound be represented mathematically?

There are very regular patterns in music, and some of them are explored mathematically here. The main mathematical idea used is that of *ratio*—ratios of lengths of strings (compared with the different sounds they make) in Sections 2 and 5; and ratios of frequencies of sounds (how *fast*, relatively, two or more strings are vibrating back and forth) in Sections 3 and 4. The central idea of the musical interval between two notes involves *ratio* rather than (numerical) *difference*, so the comparison made will be relative rather than absolute—a distinction first discussed in *Units 2* and *3*.

The title theme of this block is 'Every picture tells a story'. Although music itself is made with sounds, and so is invisible, discussing music is helped enormously by diagrams and images. You will need to pay particular attention to the assistance they offer in understanding some audible relationships in music. The diagrams provided are intended to assist that comprehension but, as you have seen on a number of occasions by now, geometric diagrams are not photographs: just like maps and graphs, they stress particular things and ignore others.

Particular images are given prominence in different sections. In Section 1, there are diagrams which represent the position of notes in a scale, while in Sections 2 and 5 there are numerous diagrams to help you visualize the relationships between notes (especially those involving ratios). In Section 3, you will see moving images produced by an oscilloscope representing sounds; these are represented further in Section 4 by graphs on the calculator screen. The appendix shows how Western musical notation uses a mix of symbols and visual display.

This is the final unit in the second block of the course and you are over half-way through MU120. It is therefore a good time to start to take stock of the ideas met in the course so far. This unit will help you look both back and look forward.

The ideas of graphing and symbolizing relationships from other units in Block B, and the computational theme of Block A, will be used and extended. For example, there will be a chance to see two more *means* at work: the *geometric* mean and the *harmonic* mean. These will enhance what you already know about the (arithmetic) mean which you met in *Unit 2*.

Looking ahead, this unit will seed ideas and experience that later units from Blocks C and D will draw upon. You will come across a reason from deep within music for looking at raising a number to a *power*, a process called *exponentiation* which is explored in depth in *Unit 12*.

You will also study the sine function, which comprises the first element in a developing library of different functions that will be explored using the calculator. In Block C, you will see a range of mathematical functions in mathematical models for different real-world phenomena. You will meet sine again in *Unit 15*, in a further exploration of the phenomenon of sound.

So as you work on this unit, be alert for themes or mathematical ideas that you have already come across recurring in familiar or different guises. In addition, look out for the various forms of 'seeing mathematically' that were outlined in *Unit 1*.

Activity 1   *Planning and reviewing progress*

Before beginning Section 1, as well as planning your study of this unit take a few minutes to review your progress so far. This might involve thinking about what you want to achieve from this unit, and how you want to develop your learning as you move towards the next block. The questions posed here are provided to give you some ideas—you may not wish to address them all, or you may have other areas that you want to review. However, the aim is for you to allocate some time as you study this unit to stop and consciously think about your work and progress.

For Blocks A and B planning and reviewing activities have been included. You have been encouraged to take time to plan your study taking into account such things as the different course components, the assessment demands and your own study preferences. Do you feel this type of activity is of value? Does it help you to keep track of your progress, be more aware of deadlines, and so on?

Are you able to monitor your work more effectively so you have a better ideal of how long you need to allocate for different tasks and what steps to take if things do not go smoothly?

For this unit, is there a particular aspect relating to your study that you want to concentrate on? For example, identifying key points from a text, using the audio component as part of learning, using technical language, making better use of diagrams and other images, or being more organized in studying?

Are you finding it straightforward or difficult to review and evaluate your own work regularly? What sources of feedback are available to help you? Do you allocate sufficient time to use such sources constructively? Is there anything that would help or improve the way you are able to evaluate your work and progress?

There are two printed response sheets for this activity.

1 Describing aspects of musical sound

Aims   The main aim of this section is to introduce some fundamental musical phenomena: 'pitches', 'notes' and 'intervals'. These are used to explore the notion of a musical scale.   ◇

1.1 A language for music

Your main activity for this subsection is listening to and working on an audiotape sequence. On the tape, you will hear a discussion about music between two people at a piano, offering a musician's and a mathematician's view of musical sound. The speakers are Christine Hodgkinson (for the musicians) and David Pimm (for the mathematicians). A piano is used in the discussion, partly because it has interesting musical—and mathematical—relationships built into it.

There is a lot to take in from this audiotape. It is denser in ideas than many audiotape bands you have listened to so far. So remember the pause/stop button on your machine. You will probably need to make quite detailed notes on the ideas discussed. In particular, you may find it helpful to draw some diagrams in order to follow what is being described. There are also some tape frames for this audio band, so you will need this text with you as you listen.

As before, remember a good way of testing *your* understanding of what you have heard is by trying to explain some of the content of this tape to someone who has not heard it.

Activity 2   *Talking about music*

Listen to band 1 of Audiotape 3.

As you listen, make a note of any new or unfamiliar words you hear and try to write a sentence or two explaining their meaning. You may want to include the following words and phrases: 'pitch' (higher and lower), 'frequency' (rate of vibration), 'tone colour' (or 'timbre'), 'names for notes', 'sharp' and 'flat', 'scale', two notes being an 'octave' apart, 'shortening violin string lengths', 'semitone', 'tone', 'pattern of notes in a scale', 'intervals', 'melody', 'harmony', 'chromatic scale'.

'Timbre' is explained more in the video band in Section 3.

You may find it useful to define these words for your handbook. If you are not able to find clear explanations of all the terms you need, do not worry, you will come across them again later in this unit, so you can add to your definitions.

There is a printed Handbook sheet for this activity.

Frame 1

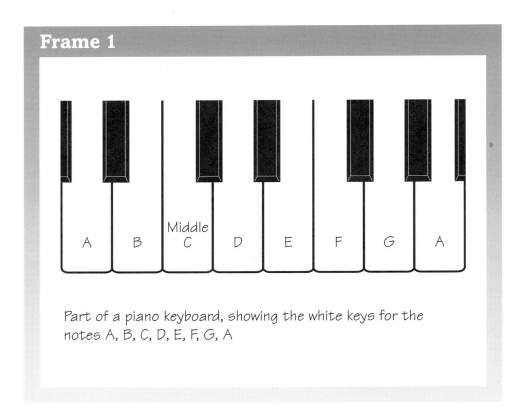

Part of a piano keyboard, showing the white keys for the notes A, B, C, D, E, F, G, A

Frame 2

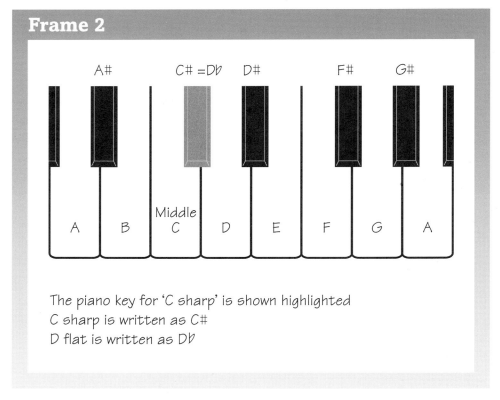

The piano key for 'C sharp' is shown highlighted
C sharp is written as C#
D flat is written as D♭

Frame 3

Frame 4

The patterns of tones and semitones in the scale of C major

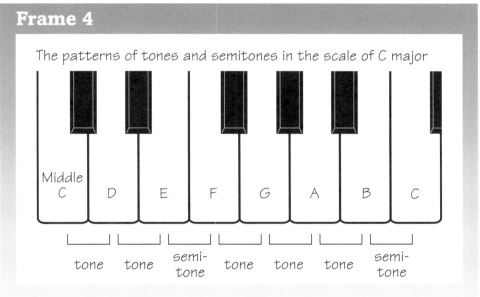

Middle C   D   E   F   G   A   B   C

tone   tone   semi-tone   tone   tone   tone   semi-tone

Activity 3   *Looking back*

Look back over the written notes you made while listening to the tape. In particular, concentrate on what you have written about scales. Are there some things you are unclear about? Write down any questions you have.

As you continue with your study, return to your questions at convenient times, to see if you can provide yourself with explanations. This will give you evidence of your developing understanding.

There is a printed response sheet on which to record your questions.

It may help to listen to parts of the tape again.

1.2   Some musical scales

One important idea presented on the audiotape was the notion of a musical scale. Your reaction to the word 'scales' may be: 'Oh, they are what musicians have to practise every day, the same patterns over and over'. But the main reason musicians concentrate on them is that in most Western music of the last three centuries, almost all of the notes in any song or piece of music will make use of a particular scale.

When a violinist slides a finger down a violin string, it produces a continuous change of pitch which passes through but does not stop to pick out individual notes. Western music picks out thirteen notes—particular pitches—in the interval known as an *octave*. These notes correspond to all the possible piano notes in one octave, using both the black and the white keys, and form what is called the *chromatic* scale.

The thirteen notes include two notes with the same letter name at the start and end of the octave interval. So the chromatic scale has twelve differently named notes in it. Each note is a semitone apart from its immediate neighbours. These are the only notes that are used in most Western music, although other cultures use different ones. So far, you have been using a diagram of a piano keyboard to show these notes. Two other ways of representing notes will be useful in understanding the relationships involved. One is a straight-line version of the piano keyboard.

The symbol ♯ is called 'a sharp' and ♭ 'a flat'.

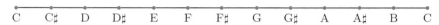

C   C♯   D   D♯   E   F   F♯   G   G♯   A   A♯   B   C

Figure 1   Line diagram of the chromatic scale

This gives equal prominence to each of the twelve differently named notes and emphasizes that the interval between any pair of consecutive notes is always the same; that is, a semitone. It is similar to the keyboard because the next octave higher in pitch will carry on to the right.

The other representation makes use of the fact that the names of the notes in the chromatic scale repeat after each octave (and there are twelve of them that have distinguishable names, like the hours on a clock).

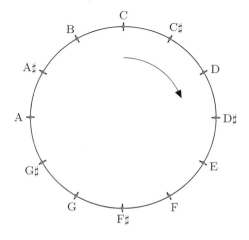

Figure 2   Circle diagram of the chromatic scale

This cyclic aspect of the notes is shown in Figure 2 by arranging them clockwise round a circle in ascending pitch order. Going up into higher octaves is indicated by going round the circle clockwise a second or third, or fourth, ... time.

Is it twelve notes or thirteen?

The two phrases 'the thirteen notes of the chromatic scale' and 'the twelve different notes of the chromatic scale' can cause confusion. It is worth spending a few moments clarifying how many notes there are.

Musicians usually speak of there being thirteen notes in an octave. For example, starting with C, these are C, C♯, D, D♯, E, F, F♯, G, G♯, A, A♯, B, C.

Probably the main reason why musicians think of an octave as having thirteen notes is that when these notes are played, to stop after the twelfth note leaves a feeling of incompleteness. It is as if you have been left hanging, waiting for the scale to be finished. Playing the thirteenth note seems to complete the scale.

However, it might seem more logical to talk about twelve notes in an octave since the thirteenth note, C again, is also the first note of a new octave. There are thus twelve differently named notes in the octave. You need to be aware of these ambiguities when you look at a picture of the thirteen keys of a piano keyboard from C to the C an octave higher, while at the same time being told that it is a representation of the twelve-note chromatic scale!

This unit sometimes refers to the chromatic scale as having thirteen notes, and sometimes as having twelve notes (meaning the twelve different notes in an octave). Often this difference does not matter, but it should be clear from the context when you are to use twelve as the number of notes.

So much for the notes. These are not to be confused with the intervals between the notes. Both mathematicians and musicians talk of there being twelve semitones in an octave: *semitone* and *octave* are names of particular intervals.

The interval called a semitone represents the step from one note in the chromatic scale to its neighbour. There are twelve such steps from C (the first note) to C (the thirteenth).

Remember the discussion on fence posts and spaces on the audiotape.

Any Western musical scale picks out a selection of notes from the full thirteen notes of the chromatic scale. This selection will be the notes used frequently in a given piece; the notes not in that scale will only be used occasionally. When a piece is composed using the notes from a single scale, it gives the piece a certain coherence and contributes to its particular character. A scale provides a musical *frame of reference* for a tune. It specifies a set of particular notes and hence draws on particular interval relationships among them. From everyday experience of music, people's ears become acclimatized to notes from different scales and they learn to recognize, and eventually expect to hear, these patterns of notes. Every culture produces musical scales.

The word 'scale' is related to the French word *échelle*, which, as well as referring to a 'musical scale', also means a 'ladder'. One way of thinking of any scale is as a 'ladder' of pitches that gets you from one starting note to the 'same' note an octave higher. A 'rung' in this ladder is then a particular musical pitch, and the gap between rungs reflects the difference in pitch, called the *musical interval* between them.

Notice that this image of a ladder and rungs has weaknesses. One is that, although a physical ladder usually has equally sized gaps between the rungs, as you have already heard, musical ladders (scales) need not.

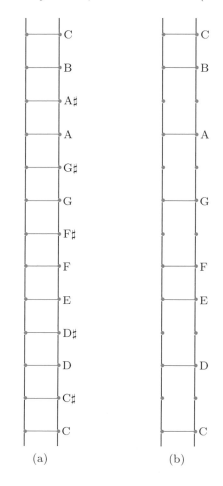

(a)   (b)

Figure 3   (a) The chromatic scale   (b) the C major scale, seen as ladders

Speaking of the major scales as having eight notes is very similar to speaking of the chromatic scale as having thirteen notes. In major scales, there are only seven *differently* named notes.

Musical scales are made up of a selection from the thirteen notes of the chromatic scale (see Figure 3). The C major and G major scales each have eight notes, as do all major scales. These eight notes include the starting note, called the *key note* (for example, 'C' here), and a note with the same name an octave higher to finish, so each major scale has seven differently named notes.

But scales in general can have any number of different notes in them, up to the limit of the complete chromatic scale of twelve different ones. The scale then repeats itself over and over again in higher (or lower) octaves. The notes in any scale are usually written from left to right in order of rising pitch.

The rest of this subsection involves activities on scale patterns. You will explore how many different scales can be created by varying the step sizes between the notes in the scale. It may well help you considerably to use the lettered piano keyboard from Frame 4 on page 11, or the circle diagram in Figure 4. The first few activities will involve uniform scales, in which the notes are selected by using a single, constant step size between all pairs of adjacent notes.

Use whichever you find more helpful.

It is important to keep clearly in mind the distinction between the actual notes in a scale, and the various step sizes *between* adjacent notes in that scale. The twelve notes which form the chromatic scale are each a semitone apart from their immediate neighbours: the scale gaps will be counted in terms of the number of semitones between adjacent notes.

Example 1   *Whole-tone scales*

One very simple rule for specifying a musical scale is to require every note to be one whole tone (that is, two semitones) away from its immediate neighbours. (Such scales are called, reasonably enough, whole-tone scales.)

(a) Is it possible to create a scale using a step size of a whole tone (that is, using a constant gap of two semitones)? Remember, the scale must start and end on the same note in a single octave.

(b) If it is possible, how many different whole-tone scales are there? (Two scales are 'the same' when all of the notes in them are identical.)

You might find it helpful to think of the chromatic scale as an underlying 'semitone scale', where each note is always one semitone away from its immediate neighbours. Bear in mind that a 'whole tone' is simply two semitones.

In thinking about these questions, one person's train of thought went something like this.

I shall use the circle diagram.

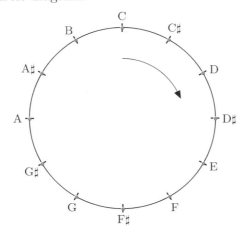

Figure 4

I have to start somewhere so I'll start at C, and mark off a whole-tone scale, stepping round the circle in whole tone (two semitones) intervals like this.

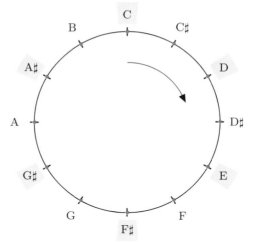

Figure 5

That's nice, it finishes exactly back on C, so it is a proper scale because it ends on the same note as it started and takes just one octave. This whole-tone scale starting on C contains the following different notes.

C, D, E, F♯, G♯, A♯

Just by looking at the circle pattern, it is clear to me that a whole-tone scale starting with any other one of the marked notes will contain exactly those same notes. For example, the whole-tone scale starting with E contains the following notes.

E, F♯, G♯, A♯, C, D

And, of course, it finishes back on E and in just one octave.

But what if I start on a note not contained in this subset—for example what if I start with the note B? I step round the circle again in whole-tone steps, but this time starting on B. The whole-tone scale then looks like the following.

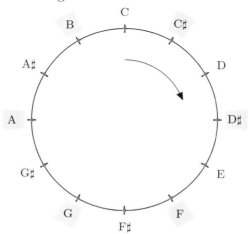

Figure 6

It finishes back at B again in one octave and so is a scale. So the whole-tone scale starting with B contains the following notes.

B, C♯, D♯, F, G, A.

As before, any whole-tone scale starting on one of these notes will produce only these notes.

Between them, these two whole-tone scales use up of all the twelve different notes in the chromatic scale. So I'm confident that there are just two different whole-tone scales.

Both the nineteenth-century classical composer Claude Debussy and the twentieth-century jazz guitarist Django Reinhardt made extensive use of whole-tone scales.

Now work on the following activities.

Activity 4 *Three semitones*

(a) How many three-semitone scales can you find? A three-semitone scale is a scale where each note is a gap of three semitones away from both its immediate neighbours. Remember each scale needs to begin and end on the same note, and span only one octave.

(b) How many different notes are there in each scale?

This sequence of activities offers an instance of how mathematicians systematically explore an idea, in an attempt to find out what is particular and special about certain situations, and what is governed by some general principle or rule.

Activity 5 *Two-tone (four semitones)*

(a) How many four-semitone scales can you find? As before, each scale needs to begin and end on the same note, and span only one octave—and each note must be four semitones away from both its neighbours.

(b) How many different notes are there in each of these scales?

The scales which are produced are *closed*. Used in this mathematical sense, 'closed' means more than that the scales merely end up on the same note that they started on. It also indicates that, for any particular scale, whatever note you start on, you will generate exactly the notes of that scale. Take, for example, the two-semitone-gap scale C, D, E, F♯, G♯, A♯, C. If you start on E using the same rule of a two-semitone gap, you will produce exactly the same set of notes.

You have now seen scales created with three different gap sizes, namely two, three and four semitones. The procedure used was similar each time. A mathematical question to ask is what is the same about these three sets of scales. Looking back over Example 1 and Activities 4 and 5 reveals some things in common between these scales. The results can be summarized in Table 1.

Table 1

Scale	Number of semitones in step (s)	Number of different notes in each scale (n)	Number of different scales (c)
Two-semitone	2	6	2
Three-semitone	3	4	3
Four-semitone	4	3	4

Collecting the results together in a table makes comparisons easy. It then can help in revealing the patterns involved. The main ones are as follows.

◇   The number of semitones in a step divides exactly into twelve, the total number of semitones in an octave. This ensures that the scale will be closed.

The other factor of twelve gives the number of notes in each scale. In symbols $s \times n = 12$.

◇   The number of notes the chromatic scale is also twelve. The result of

$12 \div$ number of notes in each scale $= 12 \div n$

$=$ the number of different scales, c.

In each case, the particular numbers are different, but the relationships among the numbers are the same. This is the sort of patterning a mathematician looks out for.

The twentieth-century German composer Arnold Schönberg was strongly associated with exploiting the chromatic scale *as a scale* (that is, drawing on *every* note equally), rather than using scales which contain only some of the notes.

If you look back at the twelve notes in the chromatic scale, you can see that the same kind of relationship exists; see Table 2.

Table 2

Scale	Number of semitones in a step	Number of different notes in each scale	Number of different scales
One-semitone (chromatic)	1	12	1

One semitone divides into twelve semitones exactly (twelve times), so this uniform scale gap size is also possible. There will be twelve different notes in the resulting scale, and there will be $12 \div 12 = 1$ such scale (which justifies calling it *the* chromatic scale).

Although these patterns arose from music, they can be described in general terms. In each case, there was a collection of things (here, twelve different notes of the chromatic scale), divided into subcollections by picking out some according to a fixed rule. The subcollections are *closed*, in that starting from any other element in the same subcollection and applying the same selection rule results in generating the same subcollection as before. Lastly, each subcollection has the same number of things in it, and the size of subcollection divides exactly into the number of things in the original collection.

So far, this method for producing scales has always worked. (The scales end on the note a single octave above the starting note.) In the next activity, you will see that this does not always happen.

Activity 6   *Will this method always produce scales?*

(a) Try to create a five-semitone scale; that is, one with a constant gap of five semitones between any two adjacent notes in it. What sequence of notes do you generate with this gap size if you start on C? What goes wrong with trying to build a five-semitone scale?

If you continue stepping in a five-semitone pattern, what sequence of notes do you get? Do you ever return to C?

(b) What about a possible six-semitone scale?

(c) Now carry out the same process with a seven-semitone scale starting on C. What sequence of notes do you generate and in what order?

The reason that five-semitone and seven-semitone 'gaps' do not produce scales is as follows. The step size, five, does not divide exactly into twelve (the number of semitones in an octave), so will not allow you to end on the same note as you started on one octave higher. Similarly, the seven-semitone gap does not produce a scale, *because* seven does not divide exactly into twelve either. The arithmetical relationships determine which scales are possible.

If you look back to your sequence of notes generated by a five- and a seven-semitone gap, you should see that notes in the two lists are identical, but in the reverse order from one another. This sequence of notes will turn up again and again, and the connection between these two orders of notes will be discussed later in this subsection.

For further investigation

As you saw, a five-semitone sequence will land on every different note in the chromatic scale—though not within the span of a single octave—before finally returning to the original starting note. For a five-semitone gap, it took five octaves to get back to C again: that is an interval of sixty semitones in all. Also, five divides exactly into sixty, twelve times, the number of notes. Similarly, for a seven-semitone gap, it takes seven octaves in all to get back to C again, and seven divides exactly into eighty-four (the number of semitones in seven octaves), twelve times, again giving the number of notes.

There appears to be another mathematical pattern here. What do you think the pattern would predict for other sized gaps, say eight semitones or eleven semitones?

Making a *conjecture* (a claim you believe to be true as a result of seeing a pattern) is an important stage in thinking mathematically about a situation. Of course, you then need to check whether your conjecture is correct!

A *chord* is a set of notes, played all together at the same time or one after another (when they are called *arpeggios*).

These examples on constant-width scale patterns should have provided you with some insight into what scales are, and also have given you a feel for how you might think about them mathematically. Equal-interval scales are not often *musically* the best choices: because of the uniform structure, no particular musical intervals stand out from one another. As was mentioned, certain composers have used whole-tone scales; the three-semitone sequence has a very important function as chords (called 'diminished seventh' chords)—but otherwise not much use is made of constant-width scales in Western music.

1.3 Building major scales

The next step is to explore scales where the gaps between adjacent notes are not a constant step size. In particular, the major scale pattern that you heard about in the audiotape is examined. Recall, from the audiotape, that the notes in the C major scale are C D E F G A B C. And in the audiotape, you were asked to work out the G major scale: G A B C D E F♯ G. The gap pattern between adjacent notes in either scale is: tone, tone, semitone, tone, tone, tone, semitone.

Scales produced using this pattern of tones and semitones are called *major scales* in Western music. These scales can be shown on circle diagrams.

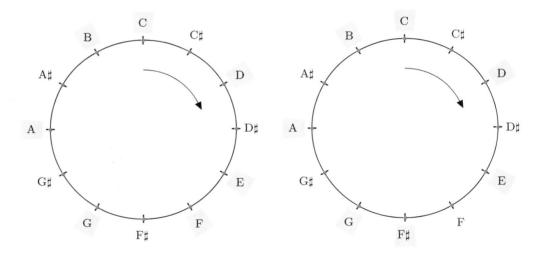

Figure 7   Circle diagrams for the C major and G major scales

You are now asked to produce some more major scales.

Activity 7   *More major scales*

(a) Start on D and use the defining pattern of tones and semitones to produce the major scale of D.

(b) In a similar way, create:

    (i) the major scale of A (starting on note A);

    (ii) the major scale of E (starting on note E).

(c) How are these five major scales of C, G, D, A, and E related to each other? What do you notice about which sharps occur in each scale?

From Activity 7, moving from the C scale → G scale → D scale → A scale → E scale, you can see that all of the notes are the same as those in the previous scale except one note is changed each time, and that note is changed up a semitone to its sharp. So, for instance, moving from the C major scale to the G major scale, the only change is that the note F is replaced by F♯. And moving from the G scale to D scale, the note C is replaced by C♯, and so on.

That is, the sharps in any new major scale generated *in this sequence* always include all of the sharps used in the previous scale. In the C scale there were no sharps; in the G scale there was one sharp (F♯); in the D scale there were two sharps (F♯ and C♯); and so on. But also look at the *position* of each new sharp in the current scale. It is always the last note before the new octave.

These patterns occur because of the sequence of step sizes in these major scales. Look at the defining sequence of steps:

    tone, tone, semitone, tone, tone, tone, semitone.

This can be abbreviated as **t**, **t**, **s**, **t**, **t**, **t**, **s**.

The way the spacing of notes begins is **t**, **t**, **s**. And this initial pattern is repeated again later in the same scale, as shown in the diagram below.

Major scale pattern of steps

Figure 8 **t**, **t**, **s** twice

This repeated block can not only explain the patterns between the scales, but also gives a method of creating major scales without having to go through the process of counting step sizes on the keyboard or on a circle diagram.

You do not need to start afresh each time you want to build a new major scale. Here is how it is done. You begin a new major scale by starting on the *fifth* note of an existing major scale.

Write out the scale for C, repeating the pattern a couple of times further.

C scale   C   D   E   F   G   A   B   C   D   E   F   G   A   B   C   D   E   F   G   ...

If you have access to a musical instrument which can produce twelve semitones to an octave, you may care to play these sequences of notes and see if you can *hear* that they have the same pattern between their respective notes, even though the particular pitches obviously differ.

Now underneath match up the G scale.

C scale	C	D	E	F	G	A	B	C	D	E	F	G	A	B	C	D	E	F	G
G scale					G	A	B	C	D	E	F♯	G	A	B	C	D	E	F♯	G

The first six notes of the G scale match those of C scale. Why? It all depends on the major scale pattern specified in terms of intervals between adjacent notes.

Note	C		D		E		F		G		A		B		C		D		E		F		G
Interval		t		t		s		t		t		t		s		t		t		s		t	

Note									G		A		B		C		D		E		F♯		G
Interval										t		t		s		t		t		t		s	

By starting a new major scale on the fifth note of the previous one, the first sequence **t**, **t**, **s** of the new scale matches the second **t**, **t**, **s** of the previous scale.

```
old    t  t  s  t  t  t  s
new             t  t  s  t  t  t  s
```

The pattern of gaps continues like this:

```
old    t  t  s  t  t  t  s  t  t | s  t | t  t  s  t  t | s  t |
new             t  t  s  t  t | t  s | t  t  s  t  t | t  s |
```

The only difference between these two patterns of step sizes occurs when the new scale has **t s** where the old scale had **s t**. This means that the notes in the two scales must be the same until this change from **s t** to **t s** occurs. Then, the new scale moves up a whole tone when the old scale moved up a semitone. In the case of the G scale, this means that the seventh note has to be F♯ and not F as it was in the C scale. The scales are brought back into step again immediately, because the new scale now moves up a semitone, but the old scale moves up a whole tone, giving a total interval gap on each scale of three semitones (whether **t s** or **s t**). So, in both the C scale and the G scale, the next note is G.

The 'sharpened' note just before the octave is called the 'leading note' of the scale.

This also explains why the change from a note to its sharp occurs on the seventh note. The difference between the pattern of step sizes for the old scale and that for the new scale occurs between the sixth and the seventh notes when there is a full tone step (**t**) in the new scale instead of the semitone step (**s**) in the old scale.

Look back to Activity 7. The note you were asked to start each scale with was the *fifth* note of the previous scale. G is the fifth note in the scale of C; D is the fifth note in the scale of G; A is the fifth note in the scale of D; E is the fifth note in the scale of A.

Activity 8   Creating further major scales

This activity uses the above method for creating scales.

(a) What will be the starting note of the next major scale after E? Produce this new scale from the scale of E major. Which note is changed to a sharp?

(b) In a similar way, create the next major scale. Which note is changed to a sharp this time?

The circle of fifths

The starting notes of this sequence of scales can be represented concisely in a diagram known as the *circle of fifths*. Part of it is shown in Fig 9.

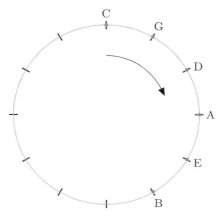

Figure 9   Part of the circle of fifths.

As you move round the circle clockwise, each note is the fifth one in the scale of the previous note. For this reason, musicians call the interval between each pair of adjacent notes 'a fifth'.

You need to be careful not to confuse this 'circle of fifths' with the circle diagram that you have been using so far. That circle diagram shows all of the notes in the chromatic scale in order going up in semitone steps—it is a chromatic scale diagram. In the circle of fifths, the step between each neighbouring pair of notes is a musical fifth.

It is quite easy to produce the circle of fifths from the chromatic scale circle diagram. To move to the fifth note in any major scale, the step sizes needed are **t**, **t**, **s**, **t**, which is seven semitones in all. So, if you count round the chromatic scale diagram in steps of seven semitones, you land on the fifth note in each major scale. Thus, for instance, from C to G is seven semitones, and so is from G to D, and also from D to A (see Figure 10).

Recall from the audiotape band that musicians usually count using only the notes in the scale and not the semitone step sizes.

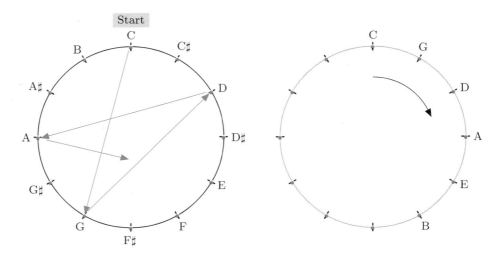

Figure 10

Activity 9   *The complete circle of fifths*

Complete the circle of fifths, checking that you eventually come back to C.

The circle of fifths connects with the work you did previously on trying to create five-semitone and seven-semitone gap scales. If you compare the order of notes around the circle of fifths with those produced by seven-semitone steps in Activity 6, you will see that they are identical:

C, G, D, A, E, B, F♯, C♯, G♯, D♯, A♯, F, C.

If you step *anti*clockwise around the circle of fifths, you produce exactly the order of notes you produced from the five-semitone steps. This is

C, F, A♯, D♯, G♯, C♯, F♯, B, E, A, D, G, C.

You need to keep a clear head here! Recall the discussion of fence posts and spaces from the audiotape.

The step size between any pair of adjacent notes is five semitones, but also each note is the *fourth* note of the previous scale. This happens because the step sizes needed to get to the fourth note are **t**, **t**, **s**, which is a total of five semitones. The musical interval between each pair of adjacent notes in this sequence is called 'a fourth'. The circle of fifths, read anticlockwise, produces the circle of fourths.

A fourth is five semitones and a fifth is seven semitones. A fourth followed by a fifth gives an octave (twelve semitones), as does a fifth followed by a fourth. So fifths and fourths are complementary to one another. Fourths, fifths and octaves are very important intervals. Discussion of combining two intervals is continued in Section 2.

If you compare the chromatic scale circle diagram and the circle of fifths, you may notice some simple relationships between the positions of the notes in the two diagrams. For example, F♯ is in the same position, but F has swapped places with B.

Different musical scales

There are numerous different scales that have been used in music from around the world. Almost all cultures recognize the phenomenon of the octave, although some choose to avoid it rather than stressing it, as happens in Western music. However, by no means all cultures divide an octave into twelve semitones. In Middle Eastern music, for example, which employs quartertones rather than semitones as the smallest musical unit, there are over three hundred different scales in use. Indian music often uses quartertones as well and some twentieth-century European composers have written music with even smaller intervals (called *microtones*).

Before the major scales started to be used in the Europe in the sixteenth century, there were a set of *modes*. Much early choral music (for example, Gregorian chants) was written in modes. One way of thinking of these modes is to imagine using only the white keys of the piano (so one mode goes, for example, from C to the C an octave above, using the white notes only: C, D, E, F, G, A, B, C). This generates a particular pattern of tones and semitones which specifies the mode. A second mode goes, for instance, from D to D using only the white notes, and so on. Each mode has a different name in the West (for instance, the pattern generated by going from C to C is called *Ionian*, and from D to D *Dorian*); whereas, in Chinese music using modes, the names vary additionally according to which note the particular mode starts on.

Even though every mode uses the same set of notes, you can perhaps see that each one will have the two semitones at different places in the sequence of notes, depending where the move from B to C and E to F occur. Only the Ionian mode, going from C to C, is the same as the major scale pattern.

There is another group of scales called pentatonic scales ('pentatonic' means 'five notes'). One such scale involves only the black notes on the piano (C♯, D♯, F♯, G♯, A♯, C♯), with a semitone gap pattern of (2, 3, 2, 2, 3). (To check whether a melody is pentatonic or not, see whether you can play it only using the black keys on a piano.) Another such scale, called a Pythagorean pentatonic scale, is C, D, E, G, A, C (2, 2, 3, 2, 3). Many traditional folk songs are either written using modes or pentatonic scales. Examples using pentatonic scales include 'Auld Lang Syne' and 'The Skye Boat Song'.

Many scales used in blues songs are related to these and usually only contain five or six different notes with the semitone gap pattern of (3, 2, 1, 1, 3, 2). They too include steps between adjacent notes larger than a tone in size. In addition to the major scales, there are also various types of *minor* scales (which are not dealt with here). Minor scales occur in three different types, each with its own pattern of tones and semitones, although they all begin 'tone, semitone'.

Activity 10   Reviewing your reviewing

By this stage of the course, you should have established a pattern of taking 'time out' to review your work and progress. But as with Activity 1 it can be worthwhile to think about *how* you are reviewing and whether or not it is a useful process for you.

For example, in this section, you will have carried out the activities in a particular order. This order may have been simply consecutive or you may have skimmed over some activities and returned to them later. Use your work on Section 1 to consider these questions:

Why did you attempt your study of Section 1 including the activities in the way you did?

Could you have used alternative ways? Do you think they would have been more or less effective? Can you summarize the main strengths and weaknesses of your work? In Activity 30, you will be asked to think about and review your use of all the different course components. You may wish to look at it now to familiarize yourself with it, or even begin to complete the printed response sheet. As you continue to use these different components, try to evaluate how you are using them.

Outcomes

After studying this section, you should be able to:

◇　explain to someone else the meaning and use of a range of musical terms including: 'names for notes', 'sharp' and 'flat', 'scale', 'two notes being an "octave" apart', 'semitone', 'tone', 'pattern of notes in a scale', 'intervals', 'chromatic scale' (Activity 2);

◇　work out equal sized gap scales and relate them to one another (Activities 4 and 5);

◇　work out what happens for gap sizes which do not produce scales (Activity 6);

◇　create major scales and explain how they relate to one another (Activities 7 and 8);

◇　understand the cycle of fifths and how it relates to the cycle of fourths (Activity 9);

◇　identify areas for improvement and devise ways to improve them (Activities 3 and 10).

2   Working with musical intervals

Aims   This section aims to show different ways of thinking about the intervals between notes.   ◇

In music, the term *interval* refers to how far apart two notes are: for example, 'the interval between these two sounds is an octave' or 'this note is an interval of a fifth above that one'. The first audiotape band introduced the notion of a musical interval and briefly mentioned the musical language for describing them.

After listening to two notes played one after another, many people can judge which is higher and which lower, but find it more difficult to say how far apart they are. The language of intervals is used to make more precise the notion of how far apart two notes are: their pitch 'distance'.

Intervals

An *interval* is the relation between two notes. The word 'interval' derives from Latin: *inter* meaning 'between' and *vallum*, meaning 'ramparts'. Hence, originally an interval was a gap to be defended. Nowadays, 'an interval' is a gap in distance or time between things, and people often speak of how *long* such an interval is. A peculiarity of the language of musical intervals is that musicians talk of the *width* rather than the length of an interval.

The usual names given to musical intervals are the words *second*, *third*, *fourth*, *fifth*, *sixth*, *seventh* and *octave*—and sometimes higher words like *ninth* or *eleventh*; for example, 'the jazz pianist Fats Waller's hand was said to span an interval of a twelfth'. It can be somewhat confusing because some of these are the same words that are used in mathematics both for naming the order of things in a list, and also for naming fractions. Their use for musical intervals is quite different.

2.1   How are musical intervals measured?

If you take the notes 'middle C' and 'G above middle C' on a piano, how far apart are they? In other words, how wide is the musical interval between them?

You have met two ways of measuring musical intervals. One is to choose a basic unit as the 'step', and count how many of these steps are required to move from C to G. The obvious step size to choose is the semitone. Counting in semitones, the notes C and G are seven semitones (or three and a half tones) apart.

The second and more common musical way of describing the size of intervals is to count using only the notes of a particular scale. The scale chosen for this purpose is the major scale which starts with the first note and contains the second note. With this method, it is the notes which are counted and not the number of *constant-width gaps* between them. In this example, G lies in the scale of C major, so count the number of *scale* notes in the scale of C major, starting with C. This gives C, D, E, F, G. As G is the fifth scale note in this sequence, G is therefore referred to as being a (musical) *fifth* above C. Likewise, E is a musical *third* above C, F a musical *fourth* above C, and so on.

The same is true in reverse. If G is a musical fifth *above* C, then we can also say that C is a musical fifth *below* G. You can check this by using the scale of G, which, in reverse, starts, G, F♯, E, D, C. The fifth note is C, and so the interval C to G is a musical fifth downwards.

You now have the opportunity to work on these ideas for yourself.

Activity 11   *Is the language of intervals consistent?*

(a) What is the interval from D up to G,

    (i) counting in semitones?

    (ii) in musical interval language?

(b) Does a drop from G to D produce the same interval descriptions as those you calculated in part (a)?

This example is intended to bring out some of the limitations of the standard interval-naming system, which is none the less part of Western cultural history.

Because both the first and last notes in a major scale have the same name, there are two different reference points from which to measure the pitch distance of the other note. So we can say that the interval between D and G is either a fifth or a fourth, depending which is above the other. So, a more accurate statement about the interval between D and G would be that G is a fourth above D and/or G is a fifth below D. And likewise that D is a fourth below G or a fifth above G.

There is a problem with the musical naming of intervals, since there are only seven notes in a major scale, and so only seven intervals can be labelled this way. However, there are twelve different semitone intervals within an octave. To overcome this problem, musicians use minor scales to label other intervals.

At a convenient time you may want to add to or check your handbook definitions.

In summary, the term 'interval' in music is used to express how 'far apart' two notes sound. As you will see in the next subsection, a useful mathematical model for representing a musical note is a length (thought of as an instrument string). The interval between two different notes can then be measured mathematically by comparing the lengths of the strings that produce them.

2.2 *Intervals represented by ratios*

Musical instruments generate sounds in different ways. Violins and cellos create notes by a string, stretched taut, being made to vibrate. Other instruments, such as a clarinet or a flute, produce sounds by a column of air being vibrated; changing notes involves altering the column length. For some instruments, such as a saxophone or an oboe, the air is vibrated by a reed; for others, such as a trumpet, it is the musician's lips and the mouthpiece that vibrate the air.

There are two basic types of stringed instruments: those like a harp where each string makes one note only, and those like a violin or cello or guitar, where the effective length of the vibrating string can be altered by the musician.

With this last type of instrument, a further distinction is worth noting. The guitar and the sitar are different from the violin and cello because they are *fretted*. Whereas a violinist or cellist can make a free choice of how to shorten the length of string, a guitarist's choices are limited by the positioning of the frets.

Guitar and cello

The central question examined in this section is:

How long do strings need to be to produce different notes?

Answering this question will involve setting up a mathematical model which connects the length of string with the note it produces when vibrated. This model concentrates only on the length of the string and ignores many other features, such as its tension, and the physical material from which it is made, all of which influence pitch. Although each of these factors affects which note is produced, the model used here assumes that they are the same for all of the strings being considered.

Recall that the 'pitch' of a note refers to how high or low it sounds.

Strings

The tension of a string and the material it is made of are crucial to the note produced by it. Strings of musical instruments are usually made from metal wire or plastic, although animal gut, often from lambs, was still used this century for violin strings. The basic note produced by a violin string is determined by altering the tension of the string, rather than by changing its length. This is what the tuning pegs do. The other notes produced by that string are then made by shortening the string length that is free to vibrate, by pressing the string down on the fingerboard.

Much of what follows, therefore, is about the relative lengths of strings, but this will translate into relationships among the sounds. Western accounts of mathematics and music look back to Pythagoras, the ancient Greek, who is claimed to have undertaken early investigations into sound relationships. Although the story connecting Pythagoras with these observations is doubtful, the discoveries themselves are true and striking. The first fact is that the pitch produced by a vibrating string goes down as the length increases. Correspondingly, the shorter the string length, the higher the pitch that is produced.

Furthermore, when the length of string is halved, the pitch of the note produced is exactly one octave above the old one. (And when the length of string is doubled, the note sounds an octave *lower* than the original one.)

What is of particular significance in these discoveries is that comparisons of string lengths were most usefully made using the *ratios* of their lengths and not the numerical *differences* between them. This distinction exactly parallels one which you have already met in *Unit 2*, when making comparisons between prices. There, the distinction was described in terms of making a relative comparison (based on the ratio of two prices) or an absolute comparison (based on their numerical difference). The advantage of using relative comparisons is that they apply regardless of the magnitude of the quantities being compared. The same advantage applies here. To sound a note an octave above a given note produced by a vibrating string, the string length must be halved. This relationship holds true regardless of the length of the original string (or of what particular note it makes).

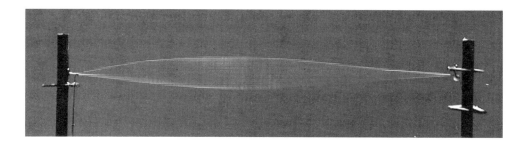

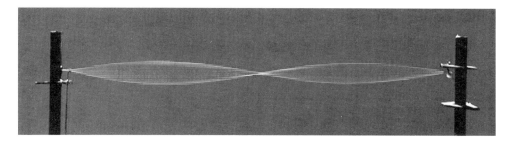

Vibrating strings giving notes an octave apart

Only one simple ratio has been mentioned so far, namely doubling or halving the length of a string to produce the note exactly an octave below or an octave above the starting note. However, the same basic principle applies whatever the ratio chosen. For example, rather than halving a string, suppose that you take three-quarters of the string length and vibrate it. Then the musical interval between the note produced by the original string and that produced by a string three-quarters its length will be the same, regardless of the particular length of the original string. It is simply the ratio between the string lengths that determines the musical interval.

Example 2   *Ratios and intervals*

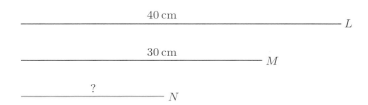

Figure 11

A string of length 40 cm is vibrated and a note is produced (string L above). The string is shortened to 30 cm (string M) and a new note produced. What new length of string is needed (string N) so that the musical interval between the notes from M and N is the same as the interval between the notes from L and M?

The string M is $\frac{30}{40} = \frac{3}{4}$ times the length of L.

For the two musical intervals to be the same, string N must also be $\frac{3}{4}$ of the length of string M.

That is, the length of string N should be $\frac{3}{4} \times 30 = 22.5\,\text{cm}$. Notice that the numerical difference between strings L and M (10 cm) and that between M and N (7.5 cm) are different.

Activity 12 *Calculating with ratios*

Using the method employed in Example 2, calculate the missing string length which would ensure that the musical interval between the notes produced by strings L and M is the same as that produced by the notes from strings M and N.

(a) String length $L\ = 9\,\text{cm}$
String length $M = 6\,\text{cm}$
String length $N\ = ?\,\text{cm}$

(b) String length $L\ = ?\,\text{cm}$
String length $M = 20\,\text{cm}$
String length $N\ = 16\,\text{cm}$

You have now seen three ways of measuring musical intervals. These are:
(a) the number of semitone steps;
(b) the number of notes counted relative to a scale;
(c) ratios of string lengths.

The connections between methods (a) and (b) were examined in Section 1. The connections between these methods and method (c), the ratios of string lengths, will be examined in this section, and in more detail in Section 5. Since ratio plays a large part, it is worth recalling the work on ratio from *Unit 2* and *Unit 6*. In those units, a ratio was used in the idea of a *multiplicative scale factor*. In other words, a given numerical value could be scaled up or down by multiplying by some constant factor. When the factor is less than 1 (say $\frac{1}{2}$ or $\frac{3}{4}$), it produces a downward scaling. Where the constant factor is numerically greater than 1 (such as $\frac{3}{2}$ or $\frac{7}{5}$), the effect of the scaling will be to increase the original value. In the next subsection, the idea of a multiplicative scale factor will be developed to investigate musical intervals further.

The use of the word 'scale' here has no connection with the musical scales described in Section 1.

2.3 *Notes from strings*

So far, lengths producing sounds an octave apart have been mentioned. The next step is to consider the notes in between. Suppose you have a string which will produce a particular note, middle C say, and a string half its length, which will produce the C an octave above. What note will the string midway in length between these two produce? Is it the note sounding midway in the octave (F♯ in this case)?

These questions are most easily answered by concentrating on the intervals between notes, rather than the notes themselves. The musical interval between the notes produced by two strings is represented by the ratio of the two string lengths.

Since the model deals only in relative lengths, there is no need to use actual string lengths; instead, to keep things simple, choose as a starting string a string of length $= 1$. The other two strings being considered will then be a string of length $= \frac{1}{2}$ (producing the note an octave above) and a third string of length $= \frac{3}{4}$ (the string exactly midway in length between the other two).

Using ratios and fractions for lengths and intervals

A ratio of one quantity to another is usually written in one of two ways:

> as a fraction: for example, $\frac{3}{4}$ (read either as 'three-quarters' or as '3 over 4');
>
> as $3:4$ (read as 'the ratio of three to four').

Writing a ratio as a fraction emphasizes that it can be treated as a multiplicative scale factor; writing it in the 'dots' form emphasizes that it can be seen as a comparison between two quantities.

Musical intervals can be seen as the ratio of two lengths of string and this ratio can be found by dividing one length by the other to produce a fraction. For example, string M, in Figure 11, is $\frac{30}{40} = \frac{3}{4}$ times the length of L. Although it would be correct to say that the musical interval between the notes produced by the two strings *is* the fraction $\frac{3}{4}$, writers on music usually say that this interval is 4:3. This can be mathematically confusing since the fraction $\frac{3}{4}$, '3 over 4', means the ratio of 3:4, and not the ratio 4:3. This confusion arises because in music, the interval is thought of as the ratio

> length of first string : length of second string.

In order to minimize the possible confusion, the mathematical form of the fraction will be used here and the 'dots' form avoided.

Recall the notion of map scale from *Unit 6*, which used the same notation.

The question to be considered is now:

> If you have a string of length $\frac{3}{4}$, it produces a certain note. What is the musical interval between this note and the starting note; and is it the same as the interval between this note and the note an octave higher than the starting note?

Example 3   *A mid-length string*

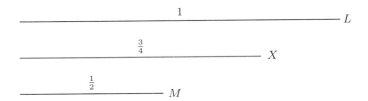

Figure 12   Three strings

String L is of length $= 1$; string M is of length $= \frac{1}{2}$; string X, halfway in-between, is of length $= \frac{3}{4}$. What are the ratios specifying the musical intervals between the notes produced by the three strings?

The interval between the notes produced by string L and string X is given by the ratio of their lengths:

$$\frac{\text{length of } X}{\text{length of } L} = \frac{\frac{3}{4}}{1} = \frac{3}{4}$$

So the musical interval from L to X is specified by the ratio $\frac{3}{4}$.

Similarly, the interval between the notes produced by string X and string M is:

$$\frac{\text{length of } M}{\text{length of } X} = \frac{\frac{1}{2}}{\frac{3}{4}} = \frac{1}{2} \times \frac{4}{3} = \frac{2}{3}$$

So the musical interval from X to M is specified by the ratio $\frac{2}{3}$.

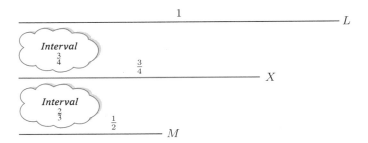

Figure 13

Since these two ratios are not the same, the musical intervals between the two pairs of notes are not equal.

Example 3 demonstrated that although the length $\left(\frac{3}{4}\right.$ of the original) was exactly halfway between the lengths giving a note and its octave, the two musical intervals on either side were not the same. So, consequently, subdividing the length exactly in half does not divide the musical interval into two equal subintervals.

Example 4 *Halving the octave*

What length of string does divide the octave exactly in half *as a musical interval?*

Call the strings which produce a note and its octave string L and string M, and the string in between (whose note will divide the octave in half as a musical interval), string X.

Then the length of the string $L = 1$, and the length of string $M = \frac{1}{2}$.

You need to find what scale factor will reduce string L to string X and then reduce string X, *using the same scale factor*, to string M, half the length of L. Call this scale factor r.

length of string $X = r \times$ (length of string L)

length of string $M = r \times$ (length of string X)

So length of string M

$$= r \times (r \times \text{length of string } L)$$
$$= r^2 \times \text{length of string } L$$
$$= r^2 = \frac{1}{2}$$

Hence $r = \sqrt{\frac{1}{2}} = \frac{1}{\sqrt{2}} = 0.7071067812$

This is approximately 0.7071 times the length of the original string. So the actual length that divides the octave into two equal intervals is 0.7071 (to four decimal places).

What does this mean in terms of actual notes? This might become clearer with an example. Suppose string L is the one that produces the note middle C, and string M produces the C an octave above this. Then, the note midway between these two will be F♯. This will be produced by a string of length $\frac{1}{\sqrt{2}} \simeq 0.7071$ times the string length that produces middle C.

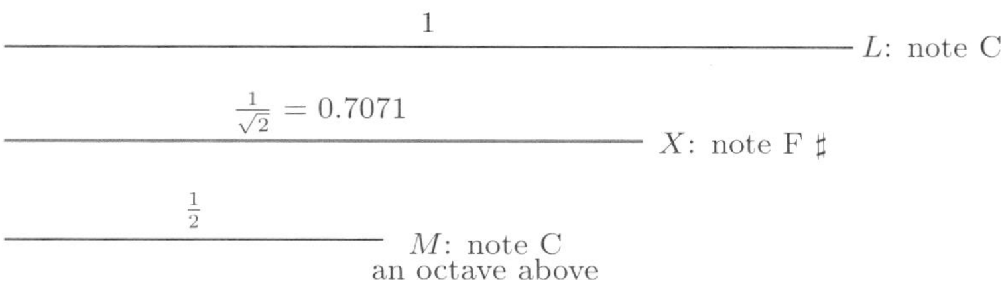

$$1 \qquad L: \text{note C}$$

$$\frac{1}{\sqrt{2}} = 0.7071 \qquad X: \text{note F } ♯$$

$$\frac{1}{2} \qquad M: \text{note C} \atop \text{an octave above}$$

Figure 14

Of course, this will not only apply to the string that produces middle C. The string of length = 0.7071 times the string length that produces the note G will produce the note C♯ which sounds midway between G and its octave. And similarly for any other string.

This interval of half an octave is known as the *tritone* (because it is three tones which equals six semitones wide). It was avoided in earlier Western music, because its rather unpleasant sound (to Western ears) became associated with 'the Devil'. It has been used more widely in the last hundred years: for example, by the Russian composer Igor Stravinsky in his ballet *Petrushka* and by the British composer Benjamin Britten in his *War Requiem*.

Activity 13   *Dividing an octave into three*

Example 4 divided an octave into two equal musical intervals. In a similar way, the octave can be divided into exactly three equal intervals, each four semitones (that is, two tones) wide (see Figure 15).

Use a method similar to that in Example 4 to:

(a) find an algebraic equation for r, the same multiplicative scale factor for each of these intervals;

(b) find the value of r, correct to four decimal places;

(c) calculate the two in-between lengths of string that will produce three equal intervals between two notes an octave apart.

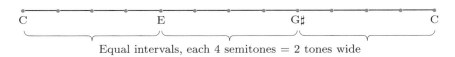

Equal intervals, each 4 semitones = 2 tones wide

Figure 15   Three equal intervals in the octave from C to C an octave above

Notice the mathematical pattern emerging:

Divide the octave into two equal intervals (each six semitones wide)	$(\text{scale factor})^2 = \frac{1}{2}$
Divide the octave into three equal intervals (each four semitones wide)	$(\text{scale factor})^3 = \frac{1}{2}$

▶   What do you expect for dividing the octave into four equal intervals, each three semitones wide? And for dividing the octave into six equal intervals, each two semitones wide?

Each string length is the length of the one before multiplied by the ratio, (or scale factor). Call it r. So the lengths of strings which divide an octave into four equal intervals will be $1, r, r^2, r^3, r^4$.

Since the last of these is the string which produces the octave, $r^4 = \frac{1}{2}$.

A similar argument holds for dividing the octave into six equal intervals (each of two semitones). The lengths of strings will be $1, r, r^2, r^3, r^4, r^5, r^6$.

Since the last of these is the string which produces the octave, $r^6 = \frac{1}{2}$.

The rest of this subsection explores the following question.

> There are twelve equal semitones in an octave. What lengths of string are needed to produce these intervals?

This is the same as asking: 'What is the ratio that represents a semitone?', because any other interval can be seen as a whole number of semitones and so can be found by multiplying the ratio for a semitone by itself the required number of times.

Activity 14   *An exact semitone*

(a) How many times does the *ratio r*, for a semitone, need to be applied to a string of length $= 1$ to get a string of length $\frac{1}{2}$?

(b) Find an algebraic equation for the ratio r, and find r to four decimal places. Explain in your own words what the value of r actually means.

The ratio r is also called the multiplicative scale factor.

Notice what this activity has shown. The musical interval of an octave can be divided equally into twelve semitones. But to do so, the interval specification involves using quite a complex decimal, and even that is still only an approximation. The chromatic scale of notes produced by this method is called an '*equally tempered*' chromatic scale. Because this ratio will give the string length to produce the semitone, it can be used to find string lengths that will produce other musical intervals—for example, thirds, fourths and fifths.

Activity 15   *Other equally tempered intervals*

In this question, r is the ratio for a semitone, the value of which you found in Activity 14.

(a) A *fifth* is seven semitones wide, so its ratio is given by r^7. What is the string length that will produce a sound a musical fifth above a given note (take its string length as 1)? How does this string length compare with $\frac{2}{3}$?

(b) A *fourth* is five semitones wide, so its ratio is given by r^5. What is the string length that will produce a sound a musical fourth above a given note? How does this string length compare with $\frac{3}{4}$?

(c) In terms of this ratio for a semitone, r, show that a fourth followed by a fifth (and also a fifth followed by a fourth) is exactly an octave.

The story of intervals does not end here. You might ask: why do we have twelve notes in an octave in the Western chromatic scale? The reasons are both historical and musical—and also mathematical—and will be examined further in Section 5. This section concludes with some mathematical writing about music from ancient Greece that gives another way of looking at some of the ideas you have just met.

2.4   On means in music

In *Unit 2*, you met the notion of 'mean' in the context of numbers and statistics. Here, you will meet three means: the *arithmetic* mean, the *geometric* mean and (in Section 4) the *harmonic* mean. These three 'means' each give a way of generating a new note (by generating a new particular length) 'in-between' any two others.

'Archytas' is pronounced 'Ah-kite-us'.

One of the first references to these means in relation to music comes from a Greek mathematician called Archytas, a follower of Pythagoras and one of Plato's teachers, who lived about 375BC. This reference occurs in one of the very few remaining fragments of his writings, one that is only eight sentences long. It may seem surprising to you that there is any hope of mathematical or musical writings from so long ago being understood. And some ideas in the Archytas extract you will shortly be looking at will probably not be immediately clear.

But you are asked to look at part of it in the next activity (and the other part in Section 4), and to make what sense of it that you can. It will then be discussed line by line and you will find that this extract refers to ideas you have already met in this section. Remember the work you did in the previous two units in this block on working with diagrams and with symbols: using symbols and diagrams may well help you here.

Recall the remark by Thompson about Cayley's working with symbols from *Unit 1*, page 18. It may offer you some inspiration.

The aim of this next activity is therefore threefold: for you to learn something more about 'means' in the context of music; for you to explore making sense of a (historical) text; and for you to see the value of using diagrams and algebraic symbolism in helping you to understand mathematical writing.

Archytas on means in music

There are three means in music. One is arithmetic, the second geometric, the third ... they call 'harmonic'.

There is an arithmetic mean when there are three terms [related] in that they exceed one another in the following way: the second exceeds the third by the same amount as that by which the first exceeds the second. In this [relation], it turns out that the musical interval between the greater terms is less and that between the lesser terms is greater.

There is a geometric mean when they are such that the [ratio of the] first is to the second, so is the second to the third. With these the interval made by the greater terms is equal to that made by the lesser.

Activity 16   *The meaning of means*

The first part of a short extract written by Archytas was given above. Read this and see what sense you can make of how each 'mean' he describes is to be calculated, and make a note of what you have difficulty following. There are two sentences in each of paragraphs 2 and 3. Which 'mean' description do you find easier to follow?

Note that when Archytas uses the word 'interval', he is referring to musical intervals exactly as have been discussed here. Words in square brackets have been added to make the original wording easier for you to follow.

Spend some time (about ten minutes) working on and thinking about the text, and only then look at the comments, and please do this before reading on.

(Note that there are two additional sentences about the harmonic mean which you will study in Section 4.)

Do not worry if you find it difficult at first.

The extract starts with two relatively straightforward sentences and then includes four more intricate ones. Number them 1 to 6. (This is a good general strategy when faced with a complex mathematical prose passage.)

Sentences 1 and 2 give the names of the three different means to be discussed, which Archytas says all relate to music. 'Harmonic' at least sounds as though it will turn out to be the most musical one of the three.

The next sentence (number 3) tells what must happen in order to have three 'terms' (numbers or lengths perhaps) being in the relation of an arithmetic mean. The relation is specified as:

> (3) the second exceeds the third by the same amount as that by which the first exceeds the second.

Think about what this is saying. When he uses the word 'exceeds' he is talking about numerical difference. Recall the work you did for Example 3, on the string $\frac{3}{4}$ times the length of the original string. This is a particular instance of what Archytas is talking about, for the three 'terms' $1, \frac{3}{4}, \frac{1}{2}$, where $\frac{3}{4}$ is the arithmetic mean of 1 and $\frac{1}{2}$.

Archytas' words can be thought about by using algebra. Introduce symbols for the three terms: x, y and z. Now there is an issue about which of x, y and z is the biggest: in the text, the first is mentioned as the biggest, so let x be the first (and biggest), y the second (middle-sized) term and z the third (and smallest) term. Figure 16 compares the situation in Example 3 and the Archytas text.

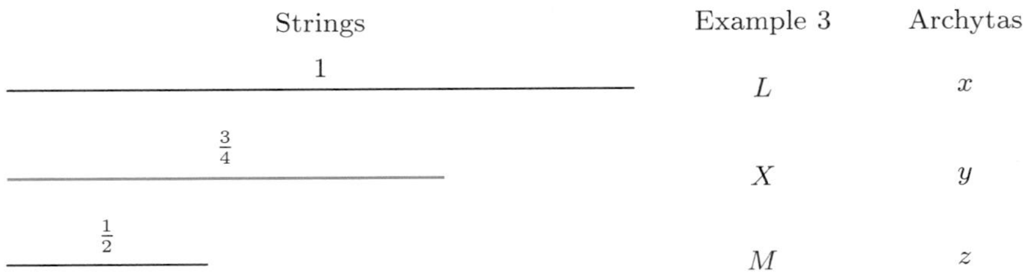

Strings	Example 3	Archytas
1	L	x
$\frac{3}{4}$	X	y
$\frac{1}{2}$	M	z

Figure 16

Translating the above verbal relationship into symbols gives:

the amount by which the second exceeds the third is $y - z$;
the amount by which the first exceeds the second is $x - y$.

So Archytas' sentence (3) claims that $y - z$ is to be the same as $x - y$. In other words,

$$y - z = x - y.$$

So y must be mid-way between x and z. Rearranging this equation by adding y to both sides (therefore not changing the equality) produces:

$$2y - z = x.$$

Adding z to both sides (again changing nothing but the *form* of the equation) produces:

$$2y = x + z.$$

Finally, in order to know how to calculate y from x and z, dividing both sides of the equation by 2 gives:

$$y = \frac{(x + z)}{2}.$$

In other words, using symbols and starting from the relationship that Archytas says must hold between three terms for them to be in the relationship of an arithmetic mean brings about a familiar formula for finding an arithmetic mean of two numbers (x and z), namely add them and divide by two.

Recall from *Unit 1*: 'Symbols can provide a way of concisely writing down what may be quite a complicated relationship if it were stated in words.'

So the algebra has helped to see what was being expressed in the words.

The middle term of the three is the arithmetic mean of the two outer terms. So now you are in a position to decide that a (for arithmetic mean) would be an even clearer choice of letter than y for the middle term. You may wish to rewrite the above calculation in your notes, substituting a for y, and thinking all the while to yourself, a is the arithmetic mean, a is what I want to find out how to calculate.

Nowadays we think of the arithmetic mean as a third number calculated from two others. The ancient Greeks seemed to think in terms of three numbers sharing a particular relationship. Nevertheless, the Greek specification can be used to tell how to calculate that middle term.

The fourth sentence is an observation about a musical property of the result:

> (4) the musical interval between the greater terms is less and that between the lesser terms is greater.

Archytas asserts that the two musical intervals are unequal, but also states which is the greater one. If you think back to the strings of length 1, $\frac{3}{4}$, $\frac{1}{2}$, you can see that this is indeed so.

As you found in Activity 15, the length of string producing an interval of a fourth (five semitones wide) is almost exactly $\frac{3}{4}$, half-way in length between 1 and $\frac{1}{2}$; and the remaining interval to the octave is then seven semitones wide, which is very close to a ratio of $\frac{2}{3}$.

A geometric model for numerical relations

Numbers can be thought of in terms of lengths of lines. This is a very common mathematical way of providing pictures of numerical relationships. In terms of a picture, imagine x and z are the lengths of line segments. Then a line whose length is the arithmetic mean $a = \dfrac{x + z}{2}$ is exactly halfway in length between x and z.

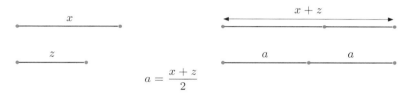

Figure 17   A diagram of the arithmetic mean

Archytas' fifth sentence, in the next paragraph, tells about calculating a geometric mean. The relationship between the terms is specified differently. But, from the above explanation of sentence 3, you might expect that once again the middle term of the three will be this new mean. So use the letters x, g (for geometric mean) and z for the three terms in order: x is again the largest of the three terms, and g is where to focus your attention.

▶   How is g to be calculated in terms of the other two?

> (5) such that the [ratio of the] first is to the second, so is the second to the third.

This is the language of ratios. The ratio of x to g is to be the same as the ratio of g to z. So expressing the relationship using symbols means:

$$\frac{x}{g} = \frac{g}{z}$$

The use of the words 'second' and 'third' to describe the terms in order and the words 'fourths' and 'fifths' as names for particular musical intervals shows again the potential for confusion between numerical and musical terminology.

Multiplying both sides of the equation first by z, and then multiplying both sides of the equation by g, gives first

$$\frac{xz}{g} = g,$$

and then

$$xz = g^2.$$

So $g^2 = xz$.

Taking square roots of both sides produces:

$$g = \sqrt{xz}$$

So, to find the geometric mean of two numbers, multiply them together and then take the square root. Once again, you have already seen one particular example of this process at work, in Example 4 on page 35.

Picturing the geometric mean

Geometrically, this has quite an interesting interpretation (as its name might suggest). If x and z are thought of as lengths rather than simply as numbers, then xz can be thought of as the area of a rectangle whose sides are of length x and z.

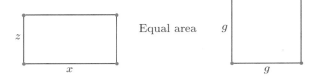

Figure 18 Diagram of the geometric mean

The relationship $g^2 = xz$ says that the area of the square whose side is g is to be the same as the area xz of the rectangle. So finding a geometric mean is equivalent to finding a square whose area is exactly that of the rectangle xz. Notice that g lies between x and z in length.

Sentence 6 is again an observation of the musical effect of this new relationship between the terms:

> (6) With these the interval made by the greater terms is equal to that made by the lesser.

To illustrate this claim further, take the sequence of lengths 24, 12, 6. With these three numbers, the middle one, 12, is the geometric mean of 24 and 6 ($12^2 = 144 = 6 \times 24$). Then the ratio of 24 to 12 is the same as the ratio of 12 to 6. Both the musical intervals formed in this particular case are the *same* width (that of an octave).

Activity 17   Looking back

Reread the Archytas extract used in Activity 16 (p. 38) and see how your understanding has changed.

Write a short commentary to demonstrate:

(a) how your understanding has changed;

(b) what has helped you in understanding the extract;

(c) what, if anything, you are still unclear or unsure about.

In summary, this section has set up a model for musical notes based on length, and explored how to find a length which produces equal musical intervals in-between two starting lengths. Generalizing the solution allowed the creation of the equally tempered chromatic scale: that is, twelve lengths all lying between a length and another half as long, where the musical intervals created between any two adjacent lengths are always the same. These intervals are each one semitone wide. The connection with finding arithmetic and geometric means was then explored, such means being about finding new lengths with particular properties in-between two given ones.

Outcomes

After studying this section, you should be able to:

◇ discuss the language of intervals in terms of semitones and terms like fourths and fifths (Activity 11);

◇ calculate intermediate string lengths between two given string lengths and decide whether the two musical intervals created are equal or not (Activity 12);

◇ find intermediate string lengths which give notes in equal musical intervals (Activities 13, 14 and 15);

◇ read and, with assistance, make sense of a historical text about means and music (Activities 16 and 17).

3 The sight of music

Aims   The main aim of this section is to introduce the idea of frequency in relation to musical pitch and to look at the notion of an equally tempered chromatic scale from this viewpoint.   ◇

Up until now, all the discussion and computation of the model for notes has been carried out in terms of lengths of strings or ratios seen as multiplicative scale factors. But you may recall that, in the first audiotape band, there was a brief mention of the idea of *frequency*, connecting it closely with musical pitch. The particular pitch of a sound is determined by the *rate of vibration*—the frequency—of the string (or column of air, or whatever causes the sound).

The first audiotape band also proposed the following question.

> Which frequencies correspond to pre-set notes on the piano and what relationships do they have to each other?

This section examines how frequencies are measured and connects them with the previous discussion of equal temperament. The main ideas are introduced in the videotape band that you will be asked to watch shortly.

As with Section 2, you may find it useful to continue to add definitions and explanations to your handbook sheet.

3.1 Frequency

A *frequency* is a number; the number of times something regular happens within a fixed period of time. With sound, the frequency is the number of vibrations made in one second. It is measured in units called *hertz* (abbreviated to Hz), where 1 Hz is 1 complete vibration per second.

A note on Hertz and physics

Heinrich Hertz was a nineteenth-century German scientist who made many measurements of features of sound and light. The station locations on your radio dial are probably measured in kHz or MHz (kiloHertz or megaHertz, that is 1000 Hz or 1 000 000 Hz, respectively). These numbers do not refer to sound wave frequencies. The medium is different (with radio waves, like other electromagnetic radiation, the signal is a variation of an electromagnetic field). This is one reason why you cannot hear radio waves directly. The human ear is sensitive to air vibrations only between approximately 20 and 20 000 Hz.

The videotape you are about to watch covers some of the same ideas that you have already worked through, but using the idea of frequency rather than concentrating on the lengths of string. It ends by posing a question concerning the construction of the guitar and its frets. Remember, the shorter the string length, the higher the pitch—and vice versa. As you will soon see, the higher the pitch, the higher the frequency number (that is, the faster the rate at which the string vibrates). So shorter strings produce sounds with higher frequencies.

Activity 18   *Video frequency*

As you watch the videotape band for this section, make notes on things that you recognize because they have already been discussed. Answer the questions below after you have watched the video.

Now watch band 1 of Videotape 2.

(a) You saw different instruments (tuning fork, violin, clarinet, flute and guitar) each play the note middle C. The traces on the oscilloscope were each different, but had a common feature. What was that feature?

(b) What was the numerical connection between the frequencies of two notes an octave apart? What was the connection between the oscilloscope traces of the two notes?

(c) When the two tuning forks were sounded together, how was the compound trace related to the traces of each fork sounded separately?

(d) What is the purpose of frets on a guitar?

There are two main strands to the videotape band you have just watched which are taken up in this unit: the idea of frequency and its representation as a trace on the oscilloscope, and the positioning of frets on a guitar. The numerical values of frequencies are looked at later in this section and a mathematical model of some oscilloscope traces is dealt with in Section 4. Constructing a guitar involves ideas you have met previously, so it is examined first.

Making a guitar

On a guitar, the string length is shortened by pressing a finger down behind a fret. This makes the effective string length that is being plucked equal to the distance between the fret and the bridge. The frets are positioned so that the notes, which are sounded by plucking the string lengths from each fret in turn, rise in the notes of the equally-tempered chromatic scale.

When guitar-makers (known as *luthiers*) fit the frets on the fingerboard, they have an overall length in which to fit roughly two octaves. On the video, this length was calibrated from 0 to 1, and one particular fret which gives the octave in the middle was marked as $\frac{1}{2}$. As you saw in Section 2,

In practice, guitar strings can vary in pitch because of differences in material, string thickness and tension. These variations seem to be more pronounced in electric guitars so these instruments usually have some mechanism built into the bridge for compensating for different strings.

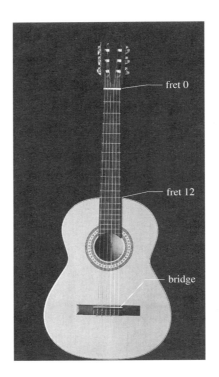

A guitar and its frets

to raise the pitch by one semitone, the string length that is plucked must be reduced by a multiplicative factor equal to $\sqrt[12]{\frac{1}{2}} = 0.9438743127$ (to 10 decimal places). If the open string (of length 1) is multiplied by that factor, it will give the position of the first fret, with distance being measured from the bridge. The second fret is produced similarly, by multiplying this new length by $\sqrt[12]{\frac{1}{2}}$, and so on. You are asked to carry out these calculations in Activity 19.

Activity 19   *Positioning the frets on a guitar*

(a) On a classical guitar there are usually 20 frets. Assuming that a guitar is calibrated from 0 to 1 (as shown in the video), use the multiplicative scale factor of $\sqrt[12]{\frac{1}{2}} = 0.9438743127$ to find the positions of the frets. Give your answers to four decimal places.

(b) The distances of the frets of an actual guitar were measured. The length of an open string, from the bridge to fret 0 was 650 mm. Calculate the distances of the frets from the bridge, and compare with the measured distances given in the comments to this activity.

Frequencies for notes

To get a string which sounds an octave above another you need to *halve* the length. More generally, the shorter the string, the higher the pitch.

With the frequencies of notes the relationship is the opposite. You were shown on the videotape band that middle C had a frequency of 256 Hz and the C an octave above (sometimes written as C′ to distinguish between the two easily) has a frequency of 512 Hz—that is, *double* that of middle C. The higher the frequency, the higher the pitch. This is described by saying that frequency is *directly* proportional to pitch, but string length is *inversely* proportional to pitch.

Each note in the chromatic scale has a particular frequency. These notes each sound a semitone above the previous one and their frequencies can be found by adapting the multiplying factor you used for producing lengths of strings in Activity 19. There, multiplying the string length by $\sqrt[12]{\frac{1}{2}}$ shortened it and raised its pitch by a semitone.

The pitch, and therefore the frequency, is inversely proportional to the length. So *reducing* the string length by a factor of $r = 0.9438743127$ *increases* the frequency by a factor of $\dfrac{1}{0.9438743127}$ or 1.059463094 (the reciprocal of r). This multiplicative scale factor is equal to the $\sqrt[12]{2}$. In other words, as the string length goes *down* by $\sqrt[12]{\frac{1}{2}}$ with each semitone, the frequency of the note goes *up* by the inverse multiplicative scale factor, $\sqrt[12]{2}$.

So the value of the ratio (to nine decimal places) that gives the increase in pitch frequency of one semitone is $s = \sqrt[12]{2} = 1.059463094$. Twelve increases of this amount raise the frequency by exactly a factor of two: that is, the pitch rises by exactly an octave. The next activity asks you to work out frequencies of the twelve notes of the equally tempered chromatic scale starting at middle C.

Directly proportional relationships were discussed in *Unit 7*. They are explored further, along with inversely proportional relationships, in *Unit 13*.

Activity 20   Frequencies of notes

Use the multiplying factor $s = \sqrt[12]{2} = 1.059463094$ to find the frequencies of each note in the chromatic scale, starting with middle C (frequency 256 Hz) and ending with the C an octave higher (frequency 512 Hz). Can you see how to use a facility on your calculator to make this computation easier?

Mathematicians and musicians describe features of frequency and pitch in rather different ways.

In the audiotape, David Pimm and Christine Hodgkinson discuss different things that struck them when they watched the videotape band that you have just watched.

In Frame 2 there is an activity which examines the difference between musical and scientific specifications of pitch.

Now listen to band 2 of Audiotape 3.

Frame 1

The ranges of notes reached by different kinds of voice

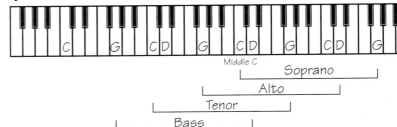

Soprano (woman): middle **C** to **G** above **C** above middle **C**

Alto : **G** below middle **C** to **D** above **C** above middle **C**

Tenor (man): **C** below middle **C** to **G** above middle **C**

Bass (man): **G** below **C** below middle **C** to **D** above middle **C**

Tenor and soprano are strong at the top of their range; and bass and alto are strong at the bottom. Thus, all notes in the vocal ranges are not covered with the same strength.
On average, men and women sing an octave apart.

Frame 2

Scientific and concert pitch

Complete the table below which shows the different frequencies of the notes of the C major scale. Give your answers to one decimal place, using the value $\sqrt[12]{2} \approx 1.059463094$
(Note that not all possible notes in the chromatic scale are listed.)

(a) Scientific pitch, for which middle **C** = 256 cycles per second
(which you have already calculated as part of Activity 20).

(b) Concert pitch, for which the A above middle C = 440 cycles per second.

Which of the listed notes differs most in absolute pitch between the two systems? Why?

	C	D	E	F	G	A	B	C
Semitone rise		2	2	1	2	2	2	1
Scientific pitch	256							512
Concert pitch						440		

Before doing this as a repeated calculation, think whether you can see how to use some of the calculator features that you have learned about in the last two units to produce the table of values in one go:
either by using the idea of a sequence generated according to a particular rule
or by a table of values arising from a particular function
or even by writing a short program.

> There are comments on this after the Comment on Activity 20

Outcomes

After studying this section, you should be able to:

◇   describe to someone else the meaning of the following terms: 'period', 'oscillation', 'frequency', 'sine wave' (Activity 18);

◇   use the idea of equal temperament to work out the frequencies of notes and fret positions on a guitar (Activities 19 and 20);

◇   describe the difference between scientific and concert pitch and compute different values for their frequencies.

4 Modelling tuning forks

Aims   The main aims of this section are to use the calculator to introduce the sine function, to explore its properties, and to use the sine function to model oscilloscope patterns from tuning forks.   ◇

There is some further exploration of this model in *Unit 15*.

You saw, on the video in Section 3, the oscilloscope traces of sounds made by tuning forks and different musical instruments. The oscilloscope generates traces from sound vibrations in the air which are picked up by a microphone, so these traces represent the vibrations given by the tuning fork or musical instrument. A mathematical description of these traces is needed, for example, by the designers of electronic instruments which produce sounds similar to acoustic musical instruments. This section makes a start at producing such a mathematical model. Figure 19 shows the traces you saw produced by a tuning fork and a violin both sounding the same pitch, middle C.

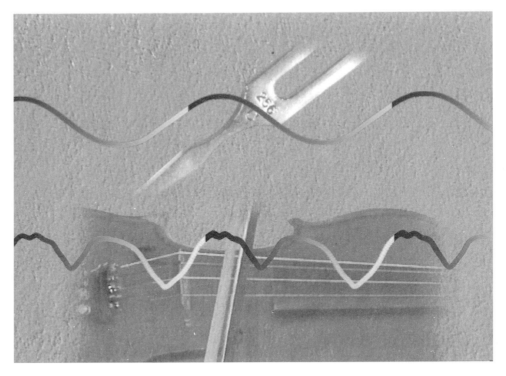

Figure 19   Traces produced by a tuning fork and a violin

The main feature of these and other sound traces is that they take the form of a wave, which can be split into identical parts. The mathematical model developed here will deal with the wave forms of only those traces created by the purest sounds—the ones produced by tuning forks. It will cover the traces made by single tuning forks and by two different tuning forks struck together, and will be used to explore the following question.

What is the acoustic effect of two different tuning forks, which vibrate at different frequencies (the number of complete vibrations a second), being struck simultaneously?

As with all mathematical models, some aspects of the situation are stressed and others ignored. One aspect that will be ignored in this model is the loudness of a note. The initial loudness of the note depends on how hard you strike the tuning fork, and that loudness eventually fades away into silence. (In the video, you saw the waves both shrink and change shape as the sound began to die away.) The model will not deal with these changes.

Whether you strike a tuning fork firmly or softly, you get the same pitch each time. The tuning fork vibrates at the same frequency, and the same shape trace is generated.

The trace produced by the sound from a tuning fork oscillates up and down and repeats itself very regularly, as you can see in Figure 19.

The oscilloscope shows a wave trace of the oscillation changing over time. It looked static on the television image, but is in fact being continuously refreshed on the screen. The videotape presenter, Ian Harrison, mentioned the term 'sine wave'. The mathematics used to describe such traces are the graphs of sine functions. Such graphs are called *sine curves*. The calculator investigation which makes up the heart of this section explores various properties of the family of sine curves.

You will see how the mathematical property known as the *period* of a sine curve (that is, the distance between repeats) is used to model frequency. Adding two different tuning fork traces (in the video, the tuning forks had frequencies of 256 and 512 Hz) produced not a sine wave, but none the less a trace which was still regularly repeating. This property is investigated by examining the period of a curve produced by adding two sine curves. In this case, one sine curve has a period twice that of the other.

Here is another example: Figure 20 shows a sketch of what you see when tuning forks of frequency 256 Hz and 384 Hz respectively are sounded together (the first frequency is two-thirds of the second): the frequency of the combined trace is 128 Hz.

The approach taken in the *Calculator Book* is to look at particular examples and then try to see what they have in common or in what ways they differ from one another. This will enable you to reach a general explanation by studying special cases.

The best way of overcoming a difficult Problem is to solve it in some particular easy cases. This gives much light into the general solution. By this way Sir Isaac Newton says he overcame the most difficult things.

(David Gregory, Scottish mathematician, 1705)

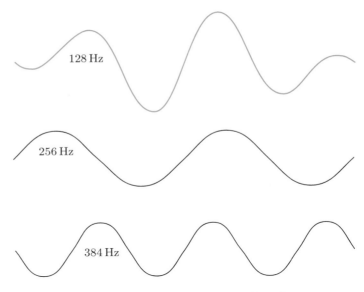

Figure 20   Traces of two tuning forks and their combined trace

In Section 9.1 of the Calculator Book, you will be asked to explore the various properties of the graphs of the trigonometric functions: sine, cosine and tangent. In Section 9.2, you will investigate the effect of adding two sine waves together.

Now work through Sections 9.1 and 9.2 of the Calculator Book.

Note that the calculator screen is not an oscilloscope. But the traces the oscilloscope generated from the sound source and the graphs of the different members of the sine family of curves are very closely linked.

4.1 Means and frequencies

In Section 2, you compared the arithmetic and geometric means in a passage by the ancient Greek mathematician Archytas. A third mean was mentioned, the *harmonic* mean. The next activity looks at the concluding paragraph of the fragment of writing by Archytas, which introduces this new mean. The harmonic mean turns out to have a surprising link with frequency.

Here are the final two sentences (numbered in accordance with the others), which talk about the harmonic mean.

More from Archytas

(7) [Then there is a mean] which we call harmonic, when they [the three terms] are such that the part of the third by which the middle term exceeds the third is the same as the part of the first by which the first exceeds the second.

(8) In this [relation], the interval between the greater terms is greater, and that between the lesser terms is less.

Activity 21 *Archytas revisited*

Re-read the first part of the Archytas extract on page 38. It will be helpful to know the following.

'Part' is an ancient Greek technical mathematical word indicating division. You can translate 'p is the part of q' as $\dfrac{p}{q}$. Also, Archytas uses 'first', 'second' and 'third' as abbreviations for 'the first term', 'the second term', and so on.

Work on this new short text in the way you did with the previous extract in Activity 16, trying to make sense of both the mathematical relation being specified and the musical claim about the intervals formed if this operation were to be carried out on string lengths.

Sentence 7 is the most complex of the whole Archytas passage. In the same way as before, now use x, h (for harmonic mean) and z for the first term, the middle term and the third term, respectively. Again, x is largest.

> (7) the part of the third by which the middle term exceeds the third is the same as the part of the first by which the first exceeds the second.

The word 'third' in the text refers to the third term (z): the middle term exceeds this by $h - z$, and this is the dividend that forms a part of z, namely $\dfrac{h - z}{z}$. The text then similarly refers to a different particular part of x, that is:

$$\frac{x - h}{x}$$

So, for a harmonic mean, you must have:

$$\frac{h - z}{z} = \frac{x - h}{x}$$

To produce a formula giving 'h = some expression involving x and z' will involve algebraic rearrangement of this equation. Algebraic manipulation of the equation $\dfrac{h - z}{z} = \dfrac{x - h}{x}$ (see box for details) gives:

$$\frac{1}{h} = \frac{1}{2}\left(\frac{1}{x} + \frac{1}{z}\right)$$

This formula can be thought of as defining the reciprocal of h (that is, $1/h$) as the arithmetic mean of the reciprocals of x and z.

Manipulating the algebra

$$\frac{h - z}{z} = \frac{x - h}{x}$$

The first step is to clear fractions. Multiply both sides by z:

$$\frac{z(h - z)}{z} = \frac{z(x - h)}{x}$$

$$h - z = \frac{z(x - h)}{x}.$$

Now multiply both sides by x, to get:

$$x(h - z) = z(x - h).$$

Multiplying out the brackets gives:

$$xh - xz = zx - zh$$

All terms involving h need to be collected together:

$$xh + zh - xz = zx$$

$$xh + zh = zx + xz \quad \text{(Note: } zx = xz\text{)}$$

and so

$$h(x + z) = 2xz$$

giving

$$h = \frac{2xz}{(x + z)} \tag{1}$$

This is a formula that will allow you to work out the value of h, if you are given x and z.

The modern formula for h is given in some books as a relationship involving $1/h$. Taking reciprocals of the formula labelled (1) above gives:

$$\frac{1}{h} = \frac{(x + z)}{2xz} = \frac{x}{2xz} + \frac{z}{2xz} = \frac{1}{2x} + \frac{1}{2z} = \frac{1}{2}\left(\frac{1}{x} + \frac{1}{z}\right)$$

The last sentence of the Archytas piece again talks about the comparative size of the musical intervals that are created, between x and h and between h and z. This can be illustrated by a specific example. If x is 12, and z is 6, then the formula specifies that h is $144/18 = 8$.

The interval ratio x to h is $\frac{2}{3}$, whereas that of h to z is $\frac{3}{4}$, and so the interval between the greater terms (x and h) in this case is indeed a wider musical interval than that between h and z, as claimed.

Catch your breath now, and then re-read the Archytas extract through. Can you see more clearly what it seems to be driving at?

Activity 22 *Putting these means in order*

It has been mentioned that each of the 'means' always lies between the two starting values. How do the different means relate to each other in size compared with a pair of fixed starting lengths x and z? One way to find out whether there is a general relationship is to look at particular cases and see if you can see something that always seems to happen.

Write down the three algebraic formulas specifying a, g and h side by side. (Formulas for a and g can be found on pages 40 and 42 respectively.) How do the values of a, g and h relate to each other for a particular pair of numbers x and z? Try some different pairs of values for x and z and see if there is something you notice that you think might happen in general.

There is one very curious thing. If you think of x and z not as numbers but as lengths of 'musical string', the earlier discussion in this section claimed that frequency of sound and length of string were inversely related. So $1/h$ is directly proportional to the frequency of the harmonic length of string, while $1/x$ is directly proportional to the frequency of the sound associated with length x, and $1/z$ directly proportional to the frequency of the sound that comes from the length z.

So one way of looking at the rewritten formula

$$\frac{1}{h} = \frac{1}{2}\left(\frac{1}{x} + \frac{1}{z}\right)$$

is that it tells us that the frequency corresponding to the length of the harmonic mean is the *arithmetic* mean of the frequencies associated with the lengths x and z. It produces a length whose frequency is in the middle of the other two.

Why is this curious? Well, one reason is that, in the fourth century BC there was no talk about frequency of vibration and little physical technology that would allow access to this inverse relationship. Yet, none the less, in music, the ancient Greeks found a way of calculating the length of a string whose frequency was the arithmetic average of the frequencies of any two other strings.

Summary of Archytas' writing

The *arithmetic* mean of two string lengths produces a new length exactly halfway between the two initial lengths (but the two musical intervals are different).

The *geometric* mean of two string lengths produces a different new length, such that the two musical intervals created are equal.

The *harmonic* mean of two string lengths produces a third new length, such that its frequency is exactly halfway between the two starting frequencies (but the two musical intervals are different).

Outcomes

After studying this section, you should be able to:

◇   recognize easily the graphs of the trigonometric functions sine, cosine and tangent;

◇   appreciate the units used for measuring angles: degrees and radians;

◇   recognize easily some of the visual properties of the family of sine curves, and use the notion of period;

◇   predict the likely period of the sum of two sine curves;

◇   compute arithmetic, geometric and harmonic means of two lengths and relate them to musical intervals (Activities 21 and 22).

5 Playing in tune

Aims   The aim of this section is to show the reasons that gave rise to the chromatic scale having twelve notes.   ◇

The starting point for this final section is the question of why there are twelve notes in the chromatic scale. Exploring the reasons for this involves examining the system of producing notes that has been used in Western music since the ancient Greeks. This not only answers the question, but also explains what it means for musical instruments to be in tune.

In Sections 1 and 2, the piano was used to introduce you to the twelve different notes of the chromatic scale and the intervals between them. This culminated in producing the ratio form for a semitone and the idea of equal temperament. In Section 3, the ratios of equally tempered notes were used to show how the frets of a guitar are placed. You may have thought the problem of how notes are created has been sorted out. But there is an important question still unanswered: why are there exactly twelve different notes in an octave?

The issue is this: a note can be produced with any frequency. You saw in Section 3 how, in scientific pitch, the note for middle C had a frequency of 256 Hz and the note for C an octave above had a frequency of 512 Hz. Between 256 and 512 an unlimited number of frequencies are possible, and, theoretically, notes for any of these values can be produced on a violin. But only twelve of these pitches are identified as notes in the chromatic scale, and pianos are designed to be able to play just these twelve different notes in any octave. How was it decided to have this number of notes? Probing this question reveals that the relationships among the notes of the chromatic scale are not as straightforward as they have appeared so far.

The present-day system of equal temperament has arisen in response to flaws in an older system of creating notes which had been used in the Western world since ancient times. This section will describe this older system—the Pythagorean system—and you will see that there are mathematical reasons for its flaws, and why equal temperament was needed to overcome them.

The problem of deciding how many notes to divide an octave into arises when different notes are played together. Pairs of notes (or more) played together can sound 'pleasant' (*harmonious* or *consonant*) or 'discordant' (*dissonant*) and one aim in producing a satisfactory chromatic scale is to have as many pairs of notes from that scale as possible sounding pleasant. But what is 'pleasant sounding'? Chords that sound pleasant to some people do not to others. Many scales within music from Eastern cultures include intervals and chords that when played can sound strange and very discordant to Western ears, and the same is true in reverse.

As well as completing this aspect of the unit you will need to complete the Learning File and handbook activities that you been working on.

In the Western musical tradition, the twelve notes of the chromatic scale are almost always the only ones identified, but Chinese and Thai music, for example, use different divisions of the octave.

Whether you hear them as pleasant or unpleasant will depend on which musical culture you have become used to. It is not only between cultures that there are these differences; within Western music, much more dissonance is now felt to be acceptable than used to be the case only a hundred years ago.

There is not space in this single unit to go into the sophistication and variation of different musical tunings and scales from around the world. In what follows, the aim is merely to explore the Western chromatic scale of twelve different notes, and how the different pitches are assigned.

5.1 Pythagoras and pleasant harmonies

In Section 2, you saw how halving the length of string produces the note an octave above. This is only one of the discoveries attributed to Pythagoras. As the scientist Jacob Bronowski put it:

> Pythagoras found a basic relation between musical harmony and mathematics. ...A single string vibrating as a whole produces a ground note. The notes that sound harmonious with it are produced by dividing the string into an exact number of parts: into exactly two parts, into exactly three parts, into exactly four parts, and so on. If the still point on the string, the node, does not come at one of these exact points, the sound is discordant.

(J. Bronowski (1973) *The Ascent of Man*, BBC Publications, p. 156)

Figure 21 shows a vibrating string divided into different numbers of parts.

If the length of the original string is 1, then the strings that sound harmonious with it are of lengths $\frac{1}{2}$, $\frac{1}{3}$, $\frac{1}{4}$ (and of course, those of complementary length $\frac{2}{3}$, $\frac{3}{4}$). The strings produced by dividing the original string into five parts, of lengths $\frac{4}{5}$, $\frac{3}{5}$, $\frac{2}{5}$, $\frac{1}{5}$ do not sound quite so well together with the original string, but were still considered to be harmonious (they are often called 'imperfect consonances', whereas the other three—$\frac{1}{2}$, $\frac{1}{3}$, $\frac{1}{4}$—are called 'perfect consonances'). This points to an essential feature of the these observations: it is dividing a string into a small number of exactly equal parts—two parts, three parts, four parts—which gives the most harmonious sounds with that of the original string. As the number of parts gets bigger, so the notes produced sound less consonant with the original string.

It is possible to use these observations to devise a method of creating new notes that will be in harmony with others. This method can be used to find new notes within an octave. Here is how it is done. Start with a string, say of length 1. The string of length $\frac{1}{2}$ (the octave above) will, as you have heard on the first audiotape band for this unit, sound pleasing played together with the string of length 1. This gives two harmonious notes. A third string of length $\frac{1}{3}$ will also produce a note that sounds pleasing when played together with the first string. However, this string produces a note outside the octave (this is because $\frac{1}{3}$ is less than $\frac{1}{2}$).

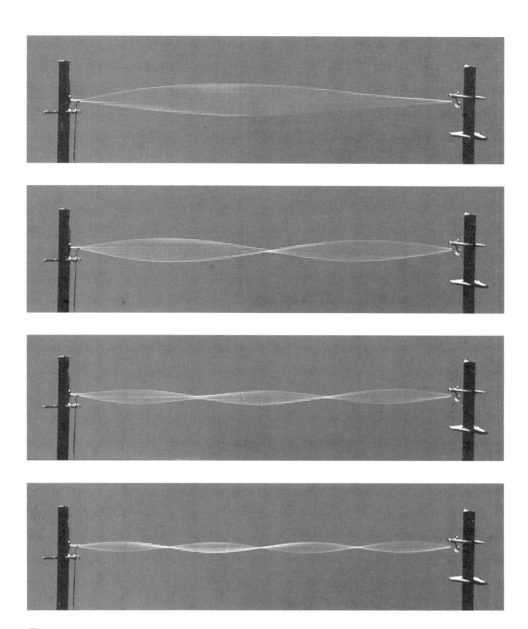

Figure 21   Vibrating strings showing different nodes

The string length which produces the same note *within* the octave is the complementary string of length $\frac{2}{3}$. So there are now three strings, of lengths 1, $\frac{1}{2}$, $\frac{2}{3}$, each producing notes within the octave and which sound harmonious. A fourth string can be produced which will sound pleasing when played with third string. This is done by taking a length that is $\frac{2}{3}$ the length of the third string. This string will be of length

$\frac{2}{3}$ of the third string $= \frac{2}{3} \times \frac{2}{3} = \frac{4}{9}$.

There are thus four notes so far, most of which sound pleasing when paired with another. The lengths of these strings can be represented on a number line in Figure 22.

Figure 22   The four strings

An important matter arises here. The strings of length

$$1, \tfrac{2}{3}, \tfrac{1}{2}$$

sound notes which lie within one octave.

But the note produced by the string of length $\frac{4}{9}$ lies in the next octave above (because $\frac{4}{9} < \frac{1}{2}$; remember $<$ means 'less than').

Since we are seeking to find the notes that lie within one octave, it seems as if this new string will not do. However, it can be replaced by the equivalent note in the octave below. This note, which of course is exactly an octave below that from the $\frac{4}{9}$ string, is obtained by doubling the length of the string to $2 \times \frac{4}{9} = \frac{8}{9}$. This string will sound just as harmonious as the string of length $\frac{4}{9}$ when played with the third string (which is of length $\frac{2}{3}$).

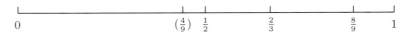

Figure 23   Notes within an octave

The procedure can be continued by creating a string which is $\frac{2}{3}$ the length of $\frac{8}{9}$ the string; that is, $\frac{2}{3} \times \frac{8}{9} = \frac{16}{27}$. The sound from this string, because its length is greater than $\frac{1}{2}$, lies within the original octave; it will sound harmonious when played together with the string of length $\frac{8}{9}$.

Activity 23   *Yet another string*

(a) Carry on this procedure to create another string. If necessary, replace it with the equivalent string within the starting octave. What is the length of this string?

(b) You now have the lengths of six strings. These were not produced in order of size. Put them in order from largest to smallest and mark the points representing the new string lengths on Figure 23. (You may find it helpful to change the fractions into decimals to do this.)

See the box on page 25.
There are now six strings with five intervals between them. The set of notes produced in this way is called a Pythagorean pentatonic scale.

Figure 24   A Pythagorean pentatonic scale

In this scale, the musical intervals between successive adjacent notes are:

Pair of notes	Interval as a fraction	Interval as decimal
$1, \frac{8}{9}$	$\frac{8}{9} \div \frac{1}{1} = \frac{8}{9}$	0.88889
$\frac{8}{9}, \frac{64}{81}$	$\frac{64}{81} \div \frac{8}{9} = \frac{8}{9}$	0.88889
$\frac{64}{81}, \frac{2}{3}$	$\frac{2}{3} \div \frac{64}{81} = \frac{27}{32}$	0.84375
$\frac{2}{3}, \frac{16}{27}$	$\frac{16}{27} \div \frac{2}{3} = \frac{8}{9}$	0.88889
$\frac{16}{27}, \frac{1}{2}$	$\frac{1}{2} \div \frac{16}{27} = \frac{27}{32}$	0.84375

There are thus two different interval sizes between adjacent pairs of notes, which fall in the pattern X, X, Y, X, Y. The pentatonic scale is used for some kinds of music, but can be rather limiting. The same process of producing new notes can be continued from the string length of $\frac{64}{81}$ to produce more notes. Calculating each of these string lengths and then putting them into order and calculating the intervals between them is rather laborious and so will be carried out using the calculator in the next subsection.

5.2   Scales on a calculator

It is worth recapping the story so far. The ancient method of producing notes for a chromatic scale, the 'Pythagorean method', exploits the remarkable discovery that, by shortening a string by a certain numerically simple ratio, the resulting musical interval between the notes produced proves to be particularly pleasing to the ear. You were then asked to generate a succession of notes by means of taking a series of such harmonious steps, each corresponding to shortening the previous string by a scale factor of $\frac{2}{3}$.

In this section, you are asked to extend these ideas, but this time supported by your calculator. You will be asked to produce another seven notes by this method and to compare these Pythagorean notes with those produced by the 'equal temperament' method used for tuning a modern piano or building a modern guitar. The equal temperament method of tuning makes every interval between successive notes of equal size (a semitone). Each of these intervals corresponds to a multiplicative scale factor of $\sqrt[12]{\frac{1}{2}}$, or 0.9438743127. The end point of this work will be to produce two lists of numbers, one corresponding to lengths of string for the 'equally tempered' chromatic scale and the other corresponding to lengths of string for the 'Pythagorean' chromatic scale.

Now work through Section 9.3 of the Calculator Book.

You have just used your calculator to generate two lists of numbers, one corresponding to lengths of string for the 'equally tempered' chromatic scale and the other corresponding to lengths of string in a 'Pythagorean' chromatic scale. Here is a reminder of how your values were produced.

Stage 1   Generating the fractions

In the calculator book, you generated the successive powers of $\frac{2}{3}$. One of the drawbacks of performing this calculation on a calculator is that the patterns in the fractions are lost (due to the fact that your calculator converted these into decimals). Table 3 below summarizes this part of the process, but it has been laid out to emphasize the fractional values of each term.

◇   Column 1 of Table 3 labels the terms 1–13.

◇   Column 2 lists successive powers of two thirds. For example, where the power of $\frac{2}{3}$ is listed as 5, the calculation required is $\left(\frac{2}{3}\right)^5$.

◇   The values of successive powers of $\frac{2}{3}$ are then listed as fractions in Column 3.

◇   Column 4 shows the scale factor by which the results of Column 3 need to be multiplied to bring them into the single octave range of 1 to $\frac{1}{2}$.

◇   Column 5 shows the results of applying the scaling, namely a list of fractions lying within the range $\frac{1}{2}$ to 1.

Table 3   The thirteen string lengths stated in fraction form

Term	Power of $\frac{2}{3}$	Value of fraction	Scale factor	Scaled fraction
1	0	1	× 1	1
2	1	$\frac{2}{3}$	× 1	$\frac{2}{3}$
3	2	$\frac{4}{9}$	× 2	$\frac{8}{9}$
4	3	$\frac{8}{27}$	× 2	$\frac{16}{27}$
5	4	$\frac{16}{81}$	× 4	$\frac{64}{81}$
6	5	$\frac{32}{243}$	× 4	$\frac{128}{243}$
7	6	$\frac{64}{729}$	× 8	$\frac{512}{729}$
8	7	$\frac{128}{2187}$	× 16	$\frac{2048}{2187}$
9	8	$\frac{256}{6561}$	× 16	$\frac{4096}{6561}$
10	9	$\frac{512}{19683}$	× 32	$\frac{16384}{19683}$
11	10	$\frac{1024}{59049}$	× 32	$\frac{32768}{59049}$
12	11	$\frac{2048}{177147}$	× 64	$\frac{131072}{177147}$
13	12	$\frac{4096}{531441}$	× 64	$\frac{262144}{531441}$

Notice, by the way, that the method as shown here is slightly different from that given in the *Calculator Book*, but essentially it amounts to the same thing. In the *Calculator Book*, you saw each newly generated power of $\frac{2}{3}$ being doubled, where necessary, at the point at which it was generated. In Table 3, the appropriate octave scalings have taken place all together at the end. This is the explanation for why the scale factor column consists of powers of 2.

Stage 2   Sorting the fractions

Table 4 is based on Table 3; but this time the values have been given as decimals. These are essentially as you saw them in the last calculator exercise, except that here they are rounded to four decimal places for convenience.

Table 4   The thirteen string lengths stated in decimal form

Term	Power of $\frac{2}{3}$	Value	Scale factor	Scaled value
1	0	1	× 1	1
2	1	0.6667	× 1	0.6667
3	2	0.4444	× 2	0.8889
4	3	0.2963	× 2	0.5926
5	4	0.1975	× 4	0.7901
6	5	0.1317	× 4	0.5267
7	6	0.0878	× 8	0.7023
8	7	0.0585	× 16	0.9364
9	8	0.0390	× 16	0.6243
10	9	0.0260	× 32	0.8324
11	10	0.0173	× 32	0.5549
12	11	0.0116	× 64	0.7399
13	12	0.0077	× 64	0.4933

Two flaws in the Pythagorean system are now evident. A guiding principle in producing the Pythagorean scale is that it should be based on *simple* numerical ratios. But the fractions produced by this method get ever more complicated and the final ones, $\left(\frac{2}{3}\right)^{11} = \frac{131072}{177147}$ and $\left(\frac{2}{3}\right)^{12} = \frac{262144}{531441}$, are far from simple!

The second flaw in the Pythagorean scale is that the final value of $\left(\frac{2}{3}\right)^{12} = \frac{262144}{531441}$ has a decimal value of 0.49327 which is close to $\frac{1}{2}$, but not exactly the same. Thus, this scale of twelve notes does not 'close on itself', which was one of the two criteria for producing scales given in Section 1.

It is possible to improve the Pythagorean scale to get more notes with smaller ratios. There is another simple fraction, $\frac{1}{4}$, which was also a perfect consonance but has not yet been used. The same method that was used for $\frac{2}{3}$ can be applied to $\frac{3}{4}$ and strings with much simpler fractions produced, at least for small powers of $\frac{3}{4}$. What you find is that the first few strings produced from $\frac{3}{4}$ are very close in length to the strings produced by the most complex fractions of $\frac{2}{3}$.

Again, because the note from a string of length $\frac{1}{4}$ falls outside the octave, the complementary length of $\frac{3}{4}$ is used.

63

Activity 24   Powers of 3/4

Calculate the values of $\frac{3}{4}$, $\left(\frac{3}{4}\right)^2$, $\left(\frac{3}{4}\right)^3$ and $\left(\frac{3}{4}\right)^4$. Scale them up by doubling, where appropriate, to create values in the range of $\frac{1}{2}$ to 1.

Convert the fractions to decimals. How do the results compare with the values of the ninth, tenth, eleventh and twelfth terms in Table 4?

The strings with the most complicated fractions can now be replaced by others. The replacement method is as follows.

◇   The final string in this sequence, with a value of 0.4933, will be replaced by the simple fraction of $\frac{1}{2}$, or 0.5. This ensures that the starting note and the thirteenth notes are exactly an octave apart.

◇   Strings 9 to 12 inclusive will be replaced by $\frac{3}{4}$, $\left(\frac{3}{4}\right)^2$, $\left(\frac{3}{4}\right)^3$ and $\left(\frac{3}{4}\right)^4$ (although not in that order, because a scaling will still have to take place to keep the fractions in the range of 0.5 to 1).

The strings with the most complicated fractions can now be replaced by others. What is left, then, is the list of string lengths shown in Table 5. The five replaced values have been written in bold to show that they were not drawn from the original sequence of powers of $\frac{2}{3}$ and their corresponding fractions are given.

The thirteen string lengths must be sorted into descending order, as shown in Table 6, to produce the amended Pythagorean chromatic scale.

Table 5

Amended terms	
1	
0.6667	$\frac{2}{3}$
0.8889	$\frac{8}{9}$
0.5926	$\frac{16}{27}$
0.7901	$\frac{64}{81}$
0.5267	$\frac{128}{243}$
0.7023	$\frac{512}{729}$
0.9364	$\frac{2048}{2187}$
0.6328	$\frac{162}{256}$
0.8438	$\frac{54}{64}$
0.5625	$\frac{9}{16}$
0.75	$\frac{3}{4}$
0.5	$\frac{1}{2}$

Table 6

Sorted terms	
1	
0.9364	$\frac{2048}{2187}$
0.8889	$\frac{8}{9}$
0.8438	$\frac{54}{64}$
0.7901	$\frac{64}{81}$
0.75	$\frac{3}{4}$
0.7023	$\frac{512}{729}$
0.6667	$\frac{2}{3}$
0.6328	$\frac{162}{256}$
0.5926	$\frac{16}{27}$
0.5625	$\frac{9}{16}$
0.5267	$\frac{128}{243}$
0.5	$\frac{1}{2}$

Recall from Section 2 that combining intervals gives $\frac{3}{4} \times \frac{2}{3} = \frac{1}{2}$.

One small thing needs to be explained. It was not merely coincidence that the small powers of $\frac{3}{4}$ replaced the most complicated fractions of $\frac{2}{3}$. This was due to the fact that musical intervals of $\frac{3}{4}$ and $\frac{2}{3}$ are complementary intervals in an octave.

So what has been produced here is a standard Pythagorean scale. In the next subsection, you will have the opportunity of comparing and contrasting this scale with the modern tuning of equal temperament.

5.3   Comparing Pythagoras with the piano

In historical terms, it should be stressed that it was the Pythagorean scale which preceded that of equal temperament. Indeed, this scale survived from the sixth century BC for about 2000 years and was used predominantly in the Western musical tradition right up to the Renaissance. Its durability over more than two millennia probably had a lot to do with the fact that it relied on certain 'perfect' proportional relationships based on very simple whole numbers. As Bronowski has remarked:

> Pythagoras had found that the chords which sound pleasing to the ear—the Western ear—correspond to exact divisions of the string by whole numbers. To the Pythagoreans this had a mystic force. The agreement between nature and number was so cogent that it persuaded them that not only the sound of nature, but all her characteristic dimensions, must be simple numbers that express harmonies. For example, Pythagoras or his followers believed that we should be able to calculate the orbits of the heavenly bodies (which the Greeks pictured as carried round the earth on crystal spheres) by relating them to the musical intervals. They felt that all the regularities in nature are musical.

> (J. Bronowski (1973) *The Ascent of Man*, BBC Publications, p. 156)

Although these simple ratios produced harmonious notes, there were serious problems with this traditional method of tuning. One flaw, which will be examined later, is that the three 'perfect' harmonies used, $\frac{1}{2}$, $\frac{2}{3}$, $\frac{3}{4}$, do not quite agree with each other. For example, as you saw in the last subsection, $\left(\frac{2}{3}\right)^{12}$ is not exactly the same as $\frac{1}{2}$, nor is $\left(\frac{3}{4}\right)^{12} = \frac{1}{2}$.

There is another substantial drawback. The intervals between successive notes, the 'Pythagorean semitones' are not all equal. This is a problem when the same melody is played together by two different instruments (or singers). Because different instruments play with different pitch ranges, the melody for one instrument needs to be transposed to a range suitable for the other. If this transposition is up or down an octave, all is well: the two instruments keep in step, an octave apart. But if the melody is transposed down some other interval, the intervals in the melody for one instrument will not be exactly the same size as those for the other. This feature of a Pythagorean scale can have excruciating consequences!

From around 1500, various attempts were made to devise other methods of tuning that allowed transportation of melodies up and down. The solution that has lasted is the method of equal temperament which began to find favour in the seventeenth century. J. S. Bach composed his famous work for the 'well tempered clavier' in 1722, in part as a celebration of this newly found flexibility of movement across different scales.

But there are inevitable costs associated with the benefits of equalizing the musical intervals by means of equally tempered tuning. The main loss is that the pitches to which notes in some modern instruments (such as the piano and guitar) are now tuned slightly 'out' in relation to what the human ear might find most harmonious. Essentially, in the interests of musical flexibility, there has been a move away from tunings which correspond to simple whole number ratios.

▶   How do the intervals in the Pythagorean and equally tempered chromatic scales compare?

The first intervals to compare are the three perfect harmonies: the intervals of $\frac{1}{2}$, $\frac{2}{3}$, $\frac{3}{4}$. The interval of $\frac{1}{2}$ corresponds to an octave and is used in both the Pythagorean and equally tempered chromatic scales. The other two perfect intervals, $\frac{2}{3}$ and $\frac{3}{4}$, correspond to the equally tempered intervals of seven semitones and five semitones, respectively. These two equally tempered intervals were discussed in Section 2 and called the intervals of 'a fifth' and 'a fourth'. For this reason, the Pythagorean interval of $\frac{2}{3}$ is called a 'perfect fifth' and the interval of $\frac{3}{4}$ is called a 'perfect fourth'. These are the actual harmonious intervals: the equally tempered intervals are only approximations to them. This is evident from Table 4 above, but you may first have noticed this in Section 2, Activity 15.

The string length values based on the two versions of the chromatic scale were calculated earlier and are repeated, rounded to four decimal places, in Table 7 for convenience. Also included are the names of the thirteen corresponding notes starting and ending on the note C.

Table 7   String length values for the equally tempered and Pythagorean scale values for the twelve-note chromatic scale starting on C

Note	Pythagorean scale	Equally tempered scale
1 = C	1.0000	1.0000
2 = C♯	0.9364	0.9439
3 = D	0.8889	0.8909
4 = D♯	0.8438	0.8409
5 = E	0.7901	0.7937
6 = F	0.7500	0.7492
7 = F♯	0.7023	0.7071
8 = G	0.6667	0.6674
9 = G♯	0.6328	0.6300
10 = A	0.5926	0.5946
11 = A♯	0.5625	0.5612
12 = B	0.5267	0.5297
13 = C	0.5000	0.5000

Activity 25   *Not so perfect intervals*

A 'perfect fifth' in music corresponds to the interval between any string and one of exactly two-thirds its length. Also, a 'perfect fourth' corresponds to the note interval between a string and one of exactly three-quarters its length.

Now use Table 7 to answer the following questions.

(a) Which note in the twelve-note scale corresponds to the note a fifth above C?

(b) Which note corresponds to the note a fourth above C?

▶ So just how close to the perfect values does the equally tempered scale place these two notes?

Comparing these intervals in the two chromatic scales may be easier if the figures are set out as in Table 8.

Table 8

Interval	Ratios corresponding to	
	Pythagorean perfect	equal temperament
A fifth	$\frac{2}{3} = 0.6667$	0.6674
A fourth	$\frac{3}{4} = 0.75$	0.7492

The calculation for the relative error of the equally tempered interval of a fourth is as follows:

$$\frac{\text{actual error}}{\text{'perfect' value}} = \frac{0.75 - 0.7492}{0.75} = 0.00107, \text{ or roughly one in a thousand.}$$

Similarly, the relative error of the equally tempered interval of a fifth is:

$$\frac{\text{actual error}}{\text{'perfect' value}} = \frac{0.6667 - 0.6674}{0.6667} = 0.0011, \text{ again roughly one in a thousand.}$$

These errors are imperceptible to most untrained ears. However, skilled musicians can detect them. When a professional violinist is playing together with a piano she may need to create slightly different notes from those made when she plays only with other stringed instruments. This is explored further in the next subsection.

How closely do the other notes correspond?

You have already seen the comparison of two particular intervals between these two scales, the fourth and the fifth, and found an extremely close correspondence (with an error of only 1 in 1000). The errors for some of the other notes are more substantial, as you will see in the next activity.

Activity 26   *Comparing the two scales*

If you have not already done so, copy the two amended lists of string lengths (corresponding to the Pythagorean and equally tempered chromatic scales) into your calculator. Take the Pythagorean scale values as representing the correct ratios.

Using a calculation similar to that on page 67, find an efficient calculator method for finding the relative error for each string length.

Pick out the three intervals which are the most 'inaccurately' represented in the equal temperament scale.

Why should a chromatic scale have any particular number of notes?

This entire section has looked at how a sequence of harmonious notes can be created within the interval of one octave. The first and last notes of this octave are represented by string lengths of 1 and $\frac{1}{2}$. As you have seen earlier the choice of five notes (the pentatonic scale) and twelve notes (the chromatic scale) were not arbitrary. Essentially, the Pythagorean story of scales is based on the fact that when you take five or twelve steps of 'perfect fifth' intervals, scaling up as you do so to stay within the octave, you find yourself *reasonably close* to the endpoint of $\frac{1}{2}$. From this point of view, one story might be that a twelve-note scale is more satisfactory than a five-note scale because the thirteenth note (value 0.4933...) is closer to 0.5 than is the sixth note (0.5267...). Additionally, the twelve-note scale offers more musical possibilities.

Perhaps you are thinking that if you keep increasing the number of notes by the same method of taking steps of perfect fifths, you might just hit the string length of $\frac{1}{2}$ *exactly*! Well, it is a nice thought, but before you launch into yet more calculation, pause and reflect a little on how these string lengths were found.

As each new string length is calculated, you multiply by $\frac{2}{3}$ and then, if necessary, multiply by two to bring it back into the desired octave range. Put another way, then, you are always multiplying the numerators of the fraction by two (or by four) and the denominators by three. This means that the numerator will always remain an even number, while the denominator will always be odd.

▶ Can this, then, ever finish up with a value of a half?

In fact, it is impossible for such a fraction (which has an even number on top and an odd number on the bottom) ever to take a value of exactly $\frac{1}{2}$. If you are not convinced of this yet, try out some equivalent fractions of $\frac{1}{2}$: for example, $\frac{3}{6}, \frac{20}{40}, \frac{25}{50}, \frac{34}{68}$. As you will soon discover, any fraction that is equivalent to $\frac{1}{2}$ must have an even denominator. Clearly, as you keep generating new fractions from multiplying by $\frac{2}{3}$, and since powers of three are all odd, this condition will never be satisfied.

There is a deeper moral to be drawn from this last piece of reflection. It is important to realize that mathematics is much more than merely carrying out calculations. Sometimes it involves taking a step back from these more mechanical aspects to ask fundamental questions such as: 'Can I ever find this?', 'Is this problem solvable?' Applying your mathematical reasoning skills may save you a lot of wasted effort in searching for the impossible, and may give you an insight into the power of mathematical thinking.

The question remains whether, rather than stop at twelve different notes in an octave, it would be possible to find other values of powers of $\frac{2}{3}$ (or $\frac{3}{4}$) that get even closer to $\frac{1}{2}$. It *is* possible, but the next power of $\frac{2}{3}$ that gets closer to $\frac{1}{2}$ than $\left(\frac{2}{3}\right)^{12}$ is $\left(\frac{2}{3}\right)^{41}$. A chromatic scale with forty-one notes within each octave, although theoretically possible, would be difficult to build into many instruments. Electronic instruments can be designed to create intervals smaller than a semitone and some composers have used scales which involve these small steps (microtones). So, one important reason why there are twelve notes in an octave is because $\left(\frac{2}{3}\right)^{12}$ is closer to $\frac{1}{2}$ than any other power of $\frac{2}{3}$ below $\left(\frac{2}{3}\right)^{41}$.

5.4   Tuning musical instruments

You may think that this discussion of perfect fifths, Pythagorean ratios and the like is only of historical interest. This is not so. Players of stringed instruments without frets play not only equally tempered intervals but can and do play perfect ones. And the two systems are both used when a piano is tuned. This may seem surprising, as you know a piano is tuned to equal temperament, but you will now hear how this is done in the concluding audiotape band for this unit.

Behind this subsection lies the question: 'What does being in tune mean?' Guitars and violins are tuned by their players, and frequently go out of tune during a performance and need to be re-tuned. A piano can be out of tune, but it is tuned by a specialist and cannot easily be altered by the pianist. 'Tuning' is making sure that the strings produce the correct note when played.

Activity 27   *Tuning a piano*

Now listen to band 3 of Audiotape 3 to hear how a piano tuner, Henry Tracy, creates perfect intervals—that is, perfect octaves, perfect fourths and perfect fifths—and then changes the perfect fourths and fifths slightly (creating intervals slightly narrower than perfect fifths, and intervals slightly wider than perfect fourths) while preserving the octaves exactly in order to achieve a good approximation to equal temperament in the course of tuning a piano. As you listen to the tape, make notes on the stages Henry Tracy went through.

Frame 1

The sequence of notes Henry tunes in the bearing scale

Remember Bb = A♯, Ab = G♯, and so on.

Start from C:

up a fourth to tune F,	then down a fifth to tune Bb
up a fourth to tune Eb,	then down a fifth to tune Ab
up a fourth to tune Db,	then down a fifth to tune Gb
up a fourth to tune B,	
then up another fourth to tune E,	then down a fifth to tune A
up a fourth to tune D,	then down a fifth to tune G

A fourth is always five semitones and a fifth always seven. The reason for the break in the pattern after B is that if you went down a fifth from B to E, you would fall outside the bearing scale.

So instead of going down a perfect fifth, Henry Tracy went up another perfect fourth to the E above this one (to a note which does lie in the bearing scale), again making use of the fact that a fourth followed by a fifth is an octave (seven semitones plus five semitones is twelve semitones, the number in an octave).

Henry Tracy with tuning lever.

Most notes are made by the hammer striking three strings.

A felt wedge blocking off one string of three for a note.

The bass notes on the far left have only one string, those more to the right, two strings, then three.

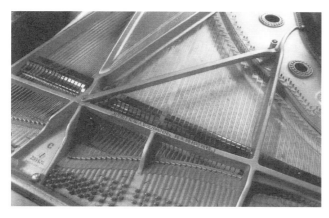

After the 'break' (where the strings go bottom right to top left rather than bottom left to top right), each note is made by threee strings being struck at once.

Felt device blocking off the outside two strings of each note in the bearing scale.

Tuning fork for the note 'C above middle C'.

Pulling the string up in pitch with the tuning lever.

Tuning pins for each string.

The bearing scale, from F♯ below middle C (his little finger) to F above middle C (his thumb).

Felt wedge blocking off two strings, leaving one free to vibrate.

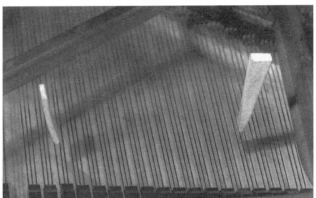

Felt wedges demarcating the bearing scale.

The next activity will ask you to review how you make summary notes.

Activity 28   Reflecting on taking notes

How easy or difficult did you find making your summary notes from someone speaking? Taking notes is an important and useful skill—but it is also important to consider why you are taking them, who they are for and how they will be used. For this activity, the notes were for yourself. How did you take them? Did you listen to the tape through once and then listen again and take notes at the same time? Did it help your concentration to take notes or was it a distraction?

As you look back on them do they now make sense? Do you think you would know what your notes referred to if you reviewed them at a later date—that is, have you labelled them clearly?

Activity 29   *Perfect fifths and octaves will not go together*

Why can you not have perfect fifths and exact octaves together on the same piano?

Outcomes

After studying this section you should be able to:

◇   calculate, with the aid of your calculator, Pythagorean scales and compare the notes produced with the equally tempered scale (using Activities 23, 24, 25, 26);

◇   describe some of the consequences of the two scales; for example, for tuning a piano (Activity 27);

◇   reflect on your own learning from different media, in particular from an audiotape (Activity 28).

Unit summary and outcomes

This unit has illustrated the way in which the Western musical scale is constructed depends on the ratios of lengths or frequencies of the notes from which it is made up. The ratios involved may be quite simple.

The mathematical ideas in this unit have been primarily about ways of counting: where to start and finish and what to include (scales and intervals), ways of combining (intervals), ways of dividing into parts (equal and Pythagorean temperaments dividing up the octave).

Another underlying theme has been about exactness and approximation. This included dividing an octave into exact, Pythagorean, intervals (perfect fourths and fifths) produced by small, whole number fractions and the approximations of the notes produced from these by other equal intervals. Equal temperament approximates these perfect intervals but divides the octave exactly into twelve equal intervals.

This unit has contained a number of different sorts of mathematical thinking. These include:

◇ *generalizing*: discovering patterns in the results when the rule for forming scales was applied to construct scales with different numbers of notes.

◇ *algebraic thinking*: unravelling what an ancient Greek mathematician/ musician wrote over 200 years ago;

◇ *mathematical argument*: arguing that however many fifths you combined, you would never reach an exact number of octaves;

◇ *approximation*: examining the approximations that lie at the heart of the Western musical chromatic scale, because of the conflict between perfect fifths and octaves;

◇ *graphical thinking*: looking at the graph of a sine function as a representation of a pure musical note, and the differences in tone associated with different instruments as the sum of related sine waves;

◇ *modelling*: representing the pitch of a note by the length of string which will produce that note when vibrated.

Tuning a piano brings together all of the aspects of music dealt with in this unit: the frequencies of a tuning fork, Pythagorean perfect fourths and fifths, and creating an equally tempered, twelve-note scale.

Western musical scales have some features that are chosen by musicians (for example, what frequency to give the pitch of middle C) and others that are fixed consequences of the mathematics (for example, the way in which the perfect harmonies do not quite fit together).

As well as the different mathematical ideas that were introduced in the unit, you have been involved in a number of different skills—particularly thinking about your own progress and communicating mathematics. At this point in the course, at the end of Block B, more than halfway through the course material, it is a good time to look at how you are studying and learning to review your progress and draw together some of the threads of the work in this block. This type of activity is sometimes called 'reflecting' and it involves being able to:

◇ analyse what you have been doing to learn;

◇ identify what you have learned and how well you have learned it;

◇ use and apply the learning to other situations.

You will be familiar with this approach from your study of the units and from the way you have been completing the activities. They have involved thinking about particular ways of doing things, such as producing summaries, using different types of diagrams, graphs, and recording these in your Learning File so that you can use these experiences to inform future work. Activity 30 asks you to apply this approach to the different course components, to consider each in turn, and to review its use.

Activity 30 *Reviewing progress*

At the beginning of this unit you were invited to review aspects of your work so far. You may find it convenient now to return to your response to this activity and Activity 10 and review your work over the second half of the course.

You can add further details to the questions posed in Activity 3. Remember to include some sort of reference, for example a date, page or section number, to chart how your ideas developed.

The final task is to review your work on technical terms from the Handbook activity.

You may like to read 'The Pythagorean Plato' in the readings. This may help you to assess how much you have learned in studying this unit.

Outcomes

After studying this unit, you should be able to:

◇ explain to someone else the meaning and use of a range of musical terms including: 'names for notes', 'sharp' and 'flat', 'scale', 'two notes being an "octave" apart', 'semitone', 'tone', 'pattern of notes in a scale', 'intervals', 'chromatic scale';

◇ work out equal sized gap scales and relate them to one another;

◇ work out what happens for gap sizes which do not produce scales;

◇ create major scales and explain how they relate to one another;

◇ understand the cycle of fifths and how it relates to the cycle of fourths;

◇ discuss the language of intervals in terms of semitones and terms like fourths and fifths;

◇ calculate intermediate lengths between two given lengths and decide whether the two musical intervals created are equal or unequal;

◇ use the construction involving taking roots to find intermediate lengths which results in equal intervals;

◇ read and, with assistance, make sense of a historical text about means and music;

◇ describe to someone else the meaning of the following terms: 'oscillation', 'frequency', 'sine wave';

◇ use the idea of equal temperament to work out the frequencies of notes and fret positions on a guitar;

◇ describe the difference between scientific and concert pitch and compute different values for their frequencies;

◇ compute the arithmetic, geometric and harmonic mean of two numbers;

◇ appreciate the units used for measuring angle;

◇ recognize the visual properties of sine, cosine and tangent curves and predict the sum of two sine curves;

◇ calculate with the aid of your calculator, Pythagorean scales and compare the notes produced with the equally tempered scale;

◇ describe some of the consequences of the two scales (for example, for tuning a piano);

◇ propose targets based upon relevant information and review targets making any necessary revision;

◇ produce and follow a schedule of learning to meet proposed targets;

◇ reflect on your own learning from different media, in particular from an audiotape.

Appendix: On musical notation

This appendix gives a short description of the way the Western musical notational system operates, and makes a link with the types of graphs drawn in *Unit 7*. Learning to read musical notation fluently can take some considerable time, just as with mathematical notation.

Musical notation makes music *visible*, offering a visual representation of musical sound. Once visible, features of sounds can be linked with features of the symbols, and those who have learned to read music know which sound to produce. Many musicians can read a musical score and *hear* the sounds in their head. Musical notation is very compressed and contains a great deal of information. It is designed for players who need to take it in at speed during a performance.

Notation provides another sense of the Block B theme of 'every picture tells a story'. Written musical notation can be seen as a graph of pitch against time. It consists of horizontal lines which show particular note pitches. Each line and each space between two adjacent lines represents a different pitch. Because most musical instruments and voices can produce only a limited range of pitches, music written for these purposes shows only the relevant part of the total possible range of pitches. Thus, music for a soprano voice will involve a different part of the framework from the music for a bass drum.

Music has to be read at speed, so it is important that musicians can find their place easily. With this in mind, the system in general use shows five continuous, closely ruled lines called a *stave* which highlights part of the background grid, like a particular form of graph paper.

'Stave' is another word for *staff*, which supports the music!

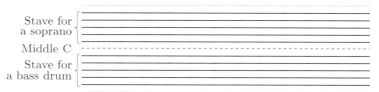

Figure 25   Stave for a soprano and stave for a bass drum

The symbol for a note is a small oval shape, and its pitch is shown by which line or space it is placed on. As Figure 25 indicates, the higher pitches are shown higher up the page. Perhaps rather surprisingly, the stave does not have a line or space for every note in the chromatic scale. Instead, only the notes A, B, C, D, E, F, G are represented by a line or a space.

The five notes of the chromatic scale that are not given lines or spaces of their own are the five notes that include sharps: A♯, C♯, D♯, F♯, G♯. These notes are indicated by putting the sign '♯' to the left of the 'unsharpened' ('natural') note symbol.

The most commonly used stave is called the 'treble stave' and is indicated by a symbol called the 'treble clef'. This is the stave shown for a soprano

Clef is the French word for key, as in the key to decoding which line is which.

singer in Figure 25 and is also used by the right hand on a piano. The bottom line of the stave represents the pitch of E above middle C.

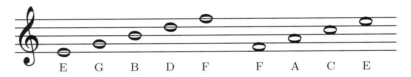

E G B D F F A C E

Figure 26 Treble stave showing treble clef and naming of notes on lines and spaces on the stave

Some people may recall mnemonic devices such as 'Every Good Boy Deserves Favour' to name successive lines of the treble stave (going upwards from the bottom line).

Activity 31 *Bass stave*

The second stave that is widely used is known as the bass stave. This is the one used by the bass drum in Figure 25 and by the left hand of a piano. The symbol for this stave is the bass clef (or 'F clef'). Where this symbol starts (the black blob) indicates which line represents the pitch of F below middle C. Work out which pitch letter is represented by each of the lines and spaces on this stave.

Figure 27

Bass and treble staves together

Piano music uses both of these staves combined to show its pitches: one stave generally for each hand. The treble stave displays music played by the right hand, and the bass stave that for the left. The image of either stave can be continued upwards or downwards by means of short extension lines which are called 'leger' lines or sometimes, particularly in North America, 'ledger' lines. '*Léger*' is a French word meaning 'light', so these can be thought of as 'lightly drawn' lines.

As Figure 28 shows, there is a gap between the two staves. An extra line between them would be needed to show middle C, and the spaces either side of this line represent the pitches B and D. There are various stories about how this particular note came to be called 'middle' C; one is that it is very near the *middle* of the piano keyboard.

Middle C

Figure 28

78

The various staves can be thought of simply as different symbols indicating the particular scale on the y-axis.

Pitch is indicated by vertical position on the stave. But what about the second axis, time duration, along the horizontal? How is this indicated? As with the writing system for English and many other Western languages, a piece of music is read from left to right, showing how notes are to follow one another in time.

The fundamental division on the horizontal axis is into *bars* by means of short vertical lines (called *bar lines*). Bar lines divide the music into time units. One important function of bar lines—just as with the stave lines—is to help align different parts in a score, so players know when they should be playing at the same instant. Bar lines are a relatively recent invention, first used in the eighteenth century. Another purpose of bar lines is to provide a regular time structure against which rhythms can be established (such as stressing the first beat in a bar—*one*, two, three, four; *one*, two, three, four; ...).

The basic rhythmical pattern of a piece of music is described by relating it to the bars: phrases such as 'waltz-time' and 'march-time' have precise meanings which are indicated by a pair of numbers at the beginning of a piece. For now, just consider one such time scale, called four-four time, written $\frac{4}{4}$. The lower number indicates the unit of time, the relative length of note given to one beat, and the upper number tells how many beats of that length are to be found in each bar.

In addition to *when* a note is to be played, the musical score shows how long a note should last.

Figure 29   A short piece of a musical score

The basic shape of a note is an oval, but within each bar, the amount of time each note is to be held for is coded into the notation for the note itself. Different note lengths are shown by being filled in or having a line with tails attached. The note length is relative to the bar unit that has been established. (Table 9 shows the relative durations of some notes.)

Finally, although the *relative* duration of notes is coded in this way, the absolute length of any note is determined by a metronome marking such as crotchet = 128. This gives the number of times per minute that a metronome would 'click', and the length of a crotchet is shown by one click.

Table 9   Relative durations, note symbols and note names

Relative time duration	Shape of note symbol	Name of note
8	‖◯‖	breve
4	◯	semibreve
2	𝅗𝅥	minim
1	𝅘𝅥	crotchet
$\frac{1}{2}$	𝅘𝅥𝅮	quaver
$\frac{1}{4}$	𝅘𝅥𝅯	semiquaver
$\frac{1}{8}$	𝅘𝅥𝅰	demi-semiquaver

Note the relative time durations are all powers of two. Can a note of any time duration always be made up by a combination of these durations?

Activity 32   *Reading music*

'Gavot' is an older spelling of the dance form usually now spelt 'gavotte'.

Look at the stave shown in Figure 30. It contains the accompanying music at the top of the page for a gavot, a type of dance, written and choreographed by Kellom Tomlinson in the year 1720. (This dance and the dance notation are illustrated in the TV programme *A Language for Movement.*)

Here are three points that may help you make sense of the music.

◇   When a note has a small dot immediately after it, the note is to last half as long again as it would normally.

◇   Individual notes can be labelled using the symbols for sharps and flats to indicate which is which, and these symbols are placed to the left of the single note to which they apply.

◇   The symbol looking like a C at the beginning stands for $\frac{4}{4}$ time.

What features do you notice? In particular, what is the highest and lowest note indicated? Are there any sharps or flats? What notes of different duration are used? How long is the relative duration of each bar, using the numbers in Table 9?

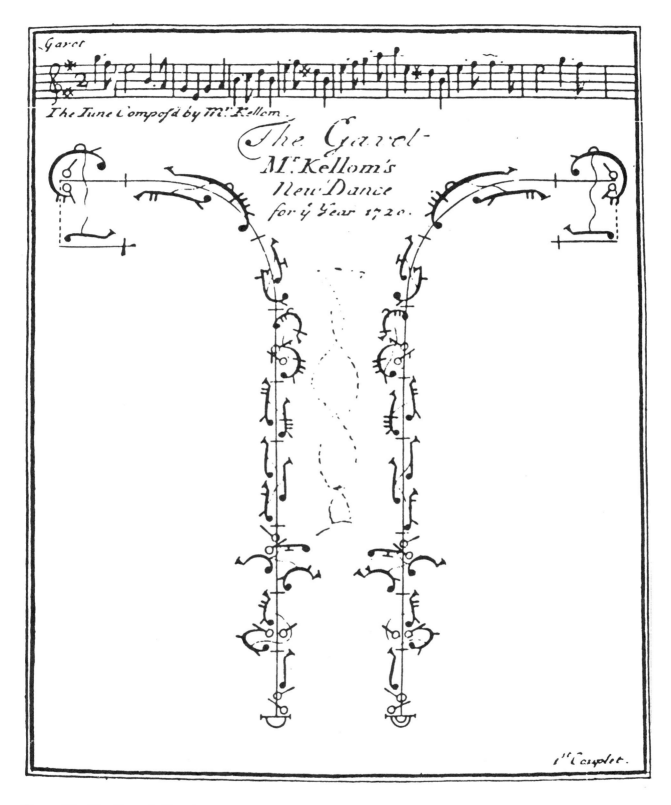

Figure 30 Tomlinson's Gavot

Comments on Activities

Activity 1

There are no comments on this activity.

Activity 2

Some of the notes one student wrote included:

> I had heard quite a lot of the words they used before, but apart from knowing 'Oh, that's a word used in music', I would have been hard pressed to say what they referred to. The difference between 'melody' and 'harmony' was one instance of this. The melody is the tune, for me, and the harmony is what goes with it.

> But having been alerted to all these words, I started to listen out for them during the day. I thought 'tone' was just another name for a note (a two-tone door chime goes 'ding-dong'), so that's one I'll have to be careful of.

> The term 'a musical fifth' was new to me, as was 'semitone'. It doesn't seem to matter that the black keys on a piano are physically smaller than the white keys, a semitone is how far apart the sounds are. I wonder what the actual mechanism is inside a piano?

The word 'chromatic' has its root in the Greek word *chromos* meaning 'colour', and gives a sense of all the different musical 'colours' available in the 'palette' provided by an instrument.

The everyday expression 'a high-pitched sound' matches the correct technical usage of the word 'pitch'.

There is an infinity of different sounds available on many instruments, such as the violin. However, human ears cannot distinguish them all, nor can human fingers produce them. The idea of 'octave', of two different pitches being fundamentally the same note, is a central idea for this unit.

Activity 3

There are no comments on this activity.

Activity 4

(a) Starting at C and stepping round the circle in steps of three semitones produces the following diagram.

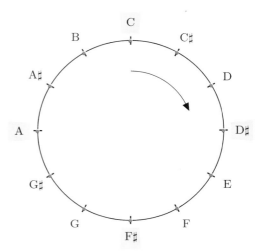

It ends back on C after one octave. This scale has four different notes in it: C, D♯, F♯, A. Starting from C♯ produces the scale C♯, E, G, A♯; and starting from D produces D, F, G♯, B.

All sets of notes are completely distinct from one another and between them contain all of the twelve different notes of the chromatic scale. There are three different three-semitone scales.

(b) Each contains four notes.

Activity 5

(a) Starting at C and stepping round the circle in steps of 4 semitones produces this diagram.

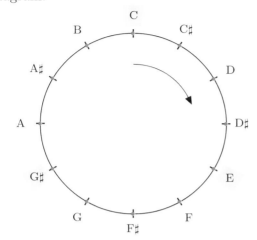

Again, it ends back on C after one octave. This scale has three different notes in it:

   C, E, G♯

There are three other scales with the notes:

   C♯, F, A;

   D, F♯, A♯;

   D♯, G, B.

There are four different four-semitone scales.

(b) Each contains three notes.

As you were doing this activity, you may have felt a sense of 'I've done this before' or 'I can begin to see what is going to happen'. Mathematicians, when they find themselves doing the 'same' thing over and over, begin to look for some common, underlying idea, that can be coded in the form of a general principle. There is more detailed discussion in the text following the activity.

Activity 6

(a) If you start to build such a scale from C, this gap-size gives the following sequence of notes: C, F, A♯. You do not land on C again and the next step takes you out of the octave. This happens because five (the step size) does not divide exactly into twelve.

If you continue stepping in five-semitone steps, the notes produced are: D♯, G♯, C♯, F♯, B, E, A, D, G, C. You have generated each note in the chromatic 'scale', although it has taken you five octaves to get back to C again.

(b) A six-semitone scale works 'properly', because six divides exactly into twelve (the number of semitones in an octave), with the answer two. So each six-semitone scale will have only two different notes in it, and there will be six different such scales. These are not very useful scales—it would be difficult to produce a tune with just two notes!

(c) With seven semitones as the step size, something very similar to the case of five semitones occurs. The sequence almost immediately goes out of the octave and eventually generates all of the notes in the full chromatic scale. This time it takes seven octaves. In order, the notes generated starting from C are:

C, G, D (now in the next octave), A, E, B, F♯, C♯, G♯, D♯, A♯, F, C. Note this is the identical list of notes in (a), simply in the reverse order.

Activity 7

(a) Work either from the keyboard chart or the circle diagram—and remember that the pattern of step size *between* adjacent notes has to be tone, tone, semitone, tone, tone, tone, semitone. Starting on D, the notes in this scale (the scale of D major) must be: D, E, F♯, G, A, B, C♯, D.

(b) (i) Starting on A results in: A, B, C♯, D, E, F♯, G♯, A.

(ii) Starting from E, the major scale pattern produces: E, F♯, G♯, A, B, C♯, D♯, E.

(c) These are the notes in some major scales.

C major   C, D, E, F, G, A, B, C . . .
G major   G, A, B, C, D, E, F♯, G . . .
D major   D, E, F♯, G, A, B, C♯, D, . . .
A major   A, B, C♯, D, E, F♯, G♯, A, . . .
E major   E, F♯, G♯, A, B, C♯, D♯, E, . . .

Each new scale starts on the fifth note of the previous one. Each scale has the same notes as the one before (including the newly created sharps), except one note is changed up a semitone. This extra sharp occurs in the penultimate position in the new scale.

Activity 8

(a) B is the fifth note in the scale of E, so the next major scale will start on B. The seventh note of the B major scale will be sharpened, so it will have to be A♯ rather than A.

(b) The fifth note of the B major scale is F♯, so the scale of F♯ major will have the same notes as B major, but will have E♯ rather than E.

(Note: E♯ is the same note as F, but the convention with major scales is to include each letter name only once in any scale.)

Activity 9

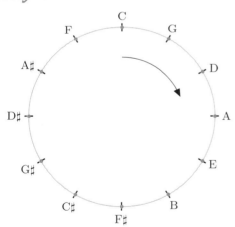

The complete cycle of fifths

Activity 10

Taking 'time out' to think about and justify why you have used one approach rather than another when you have completed a piece of work or assessment is important so you can consider what you did and whether the way you tackled it was, in retrospect, the best way.

It helps you to identify if you could improve on what you have done—and therefore help you think about how to approach other work so you can improve the quality in the future.

Activity 11

(a) (i) Counting in semitone intervals from D to G can be shown like this:

$$D \quad D♯ \quad E \quad F \quad F♯ \quad G$$
$$1 \qquad 2 \quad \ 3 \quad \ 4 \quad \ 5$$

This gives a five semitone interval.

(ii) Using musical interval notation, the interval can be deduced by looking as the major scale of D:

$$D \quad E \quad F♯ \quad G \quad A \quad B \quad C♯ \quad D.$$

Counting D as 1, E as 2, and so on, gives the result that G is a musical fourth above D.

(b) (i) Counting down in semitone intervals from G to D can be represented as in (a) (i) above. Clearly, the number of steps from D up to G (five steps) will be the same as the number of steps down from G to D. So the answer will still be five semitones.

(ii) The downward major scale of G is:

$$G \quad F♯ \quad E \quad D \quad C \quad B \quad A \quad G$$

Counting G as 1, F♯ as 2, and so on, gives the result that D is a musical fourth below G, which is the same description as before.

Activity 12

(a) $\dfrac{\text{string length } M}{\text{string length } L} = \dfrac{6}{9} = \dfrac{2}{3}.$

To maintain this interval, string length N must also be $\frac{2}{3}$ of the length of string M; that is, the length of string N should be $\frac{2}{3} \times 6 = 4\,\text{cm}.$

(b) $\dfrac{\text{string length } N}{\text{string length } M} = \dfrac{16}{20} = \dfrac{4}{5}.$

To maintain this interval, string length M must also be $\frac{4}{5}$ of the length of string L.

That is, $20\,\text{cm} = \frac{4}{5}$ of the length of string L.

So the length of string L must be $\frac{5}{4} \times 20 = 25\,\text{cm}.$

Activity 13

Call the strings L, M, N, P.

(a) The multiplicative scale factor between any string and the next (shorter) one will be the same for each pair of strings.

All the ratios multiplied together must give the musical interval between the notes produced by L and P.

(ratio from L to M) × (ratio from M to N) ×(ratio from N to P) = (ratio from L to P)

This multiplicative ratio is r. Then, since P is an octave above L, the interval L to P is $\frac{1}{2}$ and so

$$r \times r \times r = \tfrac{1}{2}.$$

(b) This simplifies to

$$r^3 = \frac{1}{2}$$

$$r = \sqrt[3]{\frac{1}{2}} = \frac{1}{\sqrt[3]{2}} = 0.7937$$

(to four decimal places).

(This value was given by the calculator.)

(c) The lengths of the strings will be as follows.
Length of string $L = 1$.
Length of string $M \simeq 0.7937$.
Length of string N
$= 0.7937 \times 0.7937 \simeq 0.6300$.
Length of string $P = 0.5$.
This interval is two tones wide. It is also the musical interval of a third from any major scale (since the major scale pattern begins **t t**; that is, two tones). So three of these thirds exactly make up an octave.

Activity 14

(a) From one semitone to the next, the string needs to be reduced by the multiplicative scale factor r. Since there are twelve semitones, the ratio needs to be applied twelve times.

(b) This means

$$r \times r \times r \times r \times r \times r \times r \times$$
$$r \times r \times r \times r \times r = \tfrac{1}{2}.$$

This can be written as:

$$r^{12} = \frac{1}{2}$$

$$r = \sqrt[12]{\frac{1}{2}} = \frac{1}{\sqrt[12]{2}} = 0.9439$$

(to four decimal places)

This value gives the musical interval of a semitone as a ratio.

Activity 15

(a) The multiplicative scale factor for a *fifth*, r^7, must be applied to the string of length 1. This gives:

$$r^7 = \left(\frac{1}{\sqrt[12]{2}}\right)^7 \simeq 0.9438743127^7 = 0.6674$$

(to four decimal places).

This is very slightly more than $\frac{2}{3} \simeq 0.6666667$.

(b) The multiplicative scale factor for a *fourth*, r^5, must be applied to the string of length 1. This gives:

$$r^5 = \left(\frac{1}{\sqrt[12]{2}}\right)^5 \simeq 0.9438743127^5 = 0.7492$$

(to four decimal places).

This is very slightly less than $\frac{3}{4} = 0.75$.

(c) A fourth followed by a fifth is
$r^7 \times r^5 = r^{12} = \frac{1}{2}$, which is exactly an octave.

A fifth followed by a fourth is
$r^5 \times r^7 = r^{12} = \frac{1}{2}$, which is also exactly an octave.

Activity 16

One student commented:

I couldn't make much of it at all at first, though once I stopped panicking and read it again I saw a little more. After the opening two sentences telling us what it is going to do, the rest is in two bits, one for each type of mean. Each bit has two sentences: one about maths and one about music. The first says something mathsy about some way of connecting the 'terms', and the second then says something about the musical 'intervals'. The details were still pretty vague for me at this stage, though.

I always think of giving a string of numbers in the order of them getting bigger. It took me a while to see that the third term was always less than the second and the second always less than the first. But the differences are said to be equal for the arithmetic one, so it sounds like the terms are like three steps in a staircase. I know from other experience that the middle one must be the average of the other two.

Archytas' piece is written in a style and a language that is not familiar nowadays. In particular, it tells about the different relationships that specify the different means all in words, with not one symbol in sight. It also talks of 'exceeds', 'terms' and 'intervals'. A much more detailed discussion is in the text following the activity.

Activity 17

There are no comments on this activity.

Activity 18

(a) All of the traces of the notes had the same number of oscillations.

(b) The frequency of the higher note was twice that of the lower note. The number of oscillations in the higher note was twice that of the lower. (So each oscillation of the higher note was only half of the length of the shorter one on the oscilloscope screen.)

(c) The compound trace had the same number of oscillations as the lower note.

(d) The purpose of the frets on a guitar is to enable the string to be shortened to produce exactly the required note.

You have seen images of sounds on the videotape oscilloscope, but the oscilloscope does not show a photograph. What you see is in effect, a graph, showing the displacement of air pressure (picked up by the microphone) plotted against time.

Activity 19

(a) The positions of the frets are shown below in the following table.

Fret number	Position
0	1
1	0.9439
2	0.8909
3	0.8409
4	0.7937
5	0.7492
6	0.7071
7	0.6674
8	0.6300
9	0.5946
10	0.5612
11	0.5297
12	0.5000
13	0.4719
14	0.4454
15	0.4204
16	0.3969
17	0.3746
18	0.3536
19	0.3337

(b) The distances of the frets from the bridge (to the nearest millimetre) are given below.

Fret number	Position	Calculated distance (mm)	Actual distance (mm)
0	1	650	650
1	0.9439	613.5	614
2	0.8909	579.1	579
3	0.8409	546.6	547
4	0.7937	515.9	516
5	0.7492	486.9	487
6	0.7071	459.6	460
7	0.6674	433.8	434
8	0.6300	409.5	409
9	0.5946	386.5	386
10	0.5612	364.8	365
11	0.5297	344.3	344
12	0.5000	325.0	325
13	0.4719	306.8	307
14	0.4454	289.5	290
15	0.4204	273.3	273
16	0.3969	258.0	258
17	0.3746	243.5	243
18	0.3536	229.8	230
19	0.3337	216.9	217

Notice that the calculated distances, rounded to the nearest millimeter are almost identical to the measured distances.

Activity 20

The frequencies of the notes (to one decimal place) in the chromatic scale starting at middle C are as follows.

Note	Frequency
C	256
C♯	271.2
D	287.4
D♯	304.4
E	322.5
F	341.7
F♯	362.0
G	383.6
G♯	406.4
A	430.5
A♯	456.1
B	483.3
C	512

Comment on Frame 2

To get equal temperament with either pitch specification, the frequency of each note has to be $\sqrt[12]{2}$ times the frequency of the note one semitone below. Some of the notes (such as D) are a whole tone above the previous one, while others (such as F or C) are only a semitone above the previously listed one (E and B respectively) in the table.

For concert pitch, the calculation is slightly trickier as 440 is in the middle of the sequence, and not at the end. But using your calculator to move backwards works well.

The values you should have obtained are shown below (to one decimal place).

Scientific pitch	C	D	E	F	G	A
	256	287.4	322.5	341.7	383.6	430.5

Scientific pitch	B	C
	483.3	512

Concert pitch	C	D	E	F	G	A
	261.6	293.7	329.6	349.2	392.0	440.0

Concert pitch	B	C
	493.9	523.3

Notice that the concert pitch values are always higher than the equivalent scientific pitch.

The differences between the two pitches for each note are as follows.

C	D	E	F	G	A	B	C
5.63	6.31	7.09	7.51	8.43	9.46	10.62	11.25

The largest difference in pitch is for the C above middle C. In fact, the difference will always rise as the notes get higher. This is because the numbers are being multiplied by the $\sqrt[12]{2}$ each time, which is a number larger than 1.

Concert pitch and scientific pitch are just two different standards which have evolved separately from the Western concert orchestra and the physics of sound.

Activity 21

There are no specific comments for this activity.

Activity 22

$$a = \frac{(x+z)}{2} \qquad g = \sqrt{xz} \qquad h = \frac{2xz}{(x+z)}$$

If you start with $x = 12$ and $z = 6$, then a is 9, $g = \sqrt{72}$ which approximately equals 8.49, and $h = 8$.

If $x = 18$ and $z = 12$, then $a = 15$, $g = \sqrt{216}$, which approximately equals 14.70, and $h = 14.4$.

While just looking at two cases clearly proves nothing in general, you should have found your values for a, g and h lying in the same order as well. $a = (x+z)/2$ is biggest, then $g = \sqrt{xz}$, then $h = 2xz/(x+z)$.

In general, it can be shown algebraically that when x is greater than z, then it is always true that $x > a > g > h > z$. In words, this says that each of the means lies between x and z, and they always occur in the same order, a bigger than g and g bigger than h.

Activity 23

(a) The length of the next string is:

$\frac{2}{3} \times \frac{16}{27} = \frac{32}{81}$. As this is less than $\frac{1}{2}$, the string length needs doubling to bring it back into the starting octave, and so
$$2 \times \frac{32}{81} = \frac{64}{81}.$$

(b) The string lengths, in the order which they were created, are $1, \frac{2}{3}, \frac{8}{9}, \frac{16}{27}, \frac{64}{81}, \frac{1}{2}$. When put in order from largest to smallest they become:

$$1, \frac{8}{9}, \frac{64}{81}, \frac{2}{3}, \frac{16}{27}, \frac{1}{2}.$$

Representation of string lengths

Activity 24

The results of the calculation are given below.

Power of $\frac{3}{4}$	Value	Scale factor	Scaled value
1	0.75	1	0.75
2	0.5625	1	0.5625
3	0.4219	2	0.8438
4	0.3164	2	0.6328

If you now compare these four values with the ones with which you wish to replace them, you can see that the correspondence is quite close. Thus:

Team	Original	replaced
9	0.6243	0.6328
10	0.8324	0.8438
11	0.5549	0.5675
12	0.7399	0.75

Activity 25

(a) The note which is a fifth above C is string number 8: the note G.

(b) The note which is a fourth above C is string number 6: the note F.

Activity 26

The calculations in the table below are based on the following formula.

Relative error =

$$\frac{\text{Equal temperament value} - \text{Pythagorean value}}{\text{Pythagorean value}}$$

The three string lengths containing the largest errors are marked **.

Equal temperament scale	Pythagorean scale	Relative error	
1	1	0	
0.9439	0.9364	0.008	**
0.8909	0.8889	0.0022	
0.8409	0.8438	−0.003	
0.7937	0.7901	0.0046	
0.7492	0.75	−0.001	
0.7071	0.7023	0.0068	**
0.6674	0.6667	0.001	
0.63	0.6328	−0.004	
0.5946	0.5926	0.0034	
0.5612	0.5625	−0.002	
0.5297	0.5267	0.0057	**
0.5	0.5	0	

These calculations indicate that if the intervals C to C♯, C to F♯ and C to B were played on a modern piano, they would represent the least satisfactory compromises that modern tuning has had to make to the Pythagorean ideal.

Activity 27

Henry starts by blocking off two of the strings that make up the note 'C above middle C' (which lies outside the bearing scale), and then tunes the third central one exactly to the pitch of the tuning fork, by adjusting the tension in the string until he hears no beats between the two sounds.

He then tunes middle C, which lies in the centre of the bearing scale, in the same way. Henry then blocks off the outside two strings of every other note in the bearing scale, and tunes the centre string of each, one by one, tuning each new note to the previously just-tuned note.

Henry then goes up a fourth from middle C to F above middle C and then from F down a fifth to B♭, then up a fourth to E♭, and so on. The only variation is where he goes up two fourths in succession, in order to stay within the octave of the bearing scale. Each time he listens for three beats between the two sounds in order to get his best approximation to equal temperament.

Henry then tunes the unisons, the other two strings which make up any note in the bearing scale. He does this with pairs of strings at a time, comparing the sound of one string with that of the already-tuned centre string, and again eliminating completely any beats between them.

So now Henry has completely tuned his reference bearing scale. He then transfers this bearing scale, starting with F♯, up and down the piano, copying a full octave note by note, before moving on to the next octave. For each note in the scale, he tunes the note an octave above (or below) it to it, again by eliminating any beats between them.

Notice that there were certain points at which Henry Tracy was free to choose what he did and other that were forced upon him by the relationships involved. For example, he chose to start with the C above middle C, and fixed that at a frequency of 523.3 Hz (that is, at concert pitch), using a tuning fork. This was not the only possible note he could have started with (nor even the only possible value of the frequency he could have given it).

Activity 28

There are no specific comments for this activity.

Activity 29

The mathematical argument on pages 68–69 is the reason why you cannot have perfect octaves and perfect fifths pre-set together on the same piano, where the mechanism for producing each note is independent of any other. And if the octaves are not exact, a piano sounds dreadful because the tuning is based upon them being the same.

What Christine Hodgkinson said about the violin is that you can tune the violin in perfect fifths and play octaves exactly, but you have to be careful to adjust your tuning (in terms of where you place your fingers to create the notes, and by avoiding playing open strings where possible) when playing music together with a piano.

Activity 30

In Activity 1 you were encouraged to think about your planning and reviewing. Studying involves both short-term and long-term planning—anything from 'studying today' to identifying how you might manage your work over the summer period and during a family holiday. Planning is all about thinking ahead, working out what needs to be done, how to go about it, and what if anything you need to complete the work.

Plans can take a variety of forms. You have been provided with a 'proforma' but you may have developed an alternative design. You may find it easier to write a plan as a set of instructions to yourself.

Obviously planning needs to be completed before you begin the actual task in hand and include thinking about such things as:

◇ what the task involves

◇ the time-scale involved for each stage of the work and the order in which each part will be done

◇ the methods/techniques to complete the work successfully

◇ the interpretation/analysis

◇ the presentation of the information.

For your study of *Unit 9* were you able to include these features?

However, a plan should be a working document. It is very rare that everything goes according to plan. Plans are only really useful if you can use them to help you check that you are meeting targets, and responding to what is happening—making changes as and when necessary. This process of monitoring your plan should help your organization of your work.

Planning and monitoring should prove valuable when you have specific deadlines to meet, for example, TMAs. However, if you find you are not able to meet your targets and things do not work out as anticipated, you need to try and identify what you need to change and perhaps seek help.

Activity 31

The following figure shows the pitch letters on the bass stave.

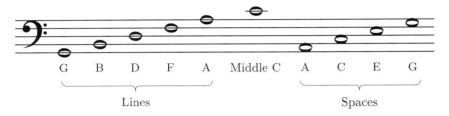

Bass stave with letters marked

Activity 32

One non-musician wrote in response:

> It is written using the treble stave. There is one note which is sharpened, the D, which appears twice (but the first time D appeared it wasn't sharpened) and none that are to be flattened. However, there were two more of these sharp symbols right at the beginning, marking both the F lines as #. Given that all the other things at the beginning applied for the whole piece of music, that may mean that all the Fs are to become F sharps (so by marking it at the beginning it has been got over with once and for all).

The musical range is from highest note of B and the lowest note shown that of E an octave and a fifth below that.

The shortest note is a quaver and the longest note is a minim, which lasts four times longer. Apart from the very first two notes, each bar has a total count of four (remember the dots). The unit is the crotchet. For instance, the bar which starts with the minim has a total count of $2 + 1\frac{1}{2}$ (because of the dot) $+ \frac{1}{2} = 4$; the next $1 + 1 + 1 + 1 = 4$; and so on.

Acknowledgements

Text

Auden, W. H. (1966) *Collected Shorter Poems 1927–1957*, Faber and Faber, © W. H. Auden, also by permission of Random House, Inc.

Cover

Train: Camera Press; map: reproduced from the 1995 Ordnance Survey 1:25 000 Outdoor Leisure Map with the permission of the Controller of Her Majesty's Stationery Office © Crown Copyright; other photographs: Mike Levers, Photographic Department, The Open University.

Index